FORTSCHRITTE DER PHYSIK

Volume 33 · 1985
Number 9

PROGRESS OF PHYSICS

CONTENTS:

AKADEMIE-VERLAG · BERLIN

ISSN 0015-8208 Fortschr. Phys., Berlin 33 (1985) 9, 495–546 EVP 10,– M

Instructions to Authors

1. Only papers not published and not submitted for publication elsewhere will be accepted.
2. Manuscripts should be submitted in English, with an abstract in English. Two copies are desired.
3. Manuscripts should be no less than 30 and preferably no more than about 100 pages in length.
4. All manuscripts should be typewritten on one side only, double-spaced and with a margin 4 cm wide. Manuscript sheets should be numerated consecutively from "1" onwards. Footnotes should be avoided.
5. The titel of the paper should be followed by the authors name (with first name abbreviated), by the institution and its address from which the manuscript originates.
6. Figures and tables should be restricted to the minimum needed to clarify the text. They should be numbered consecutively and must be referred too in the text and on the margin. Figures and tables should be added to the manuscript on separate, consecutively numerated sheets. The tables should have a headline. Legends of figures should be submitted on a separate sheet. All figures should bear the author's name and number of figure overleaf. Photographs for half-tone reproduction should be in the form of highly glazed prints. Line drawings should be in a form suitable for reproduction. The lettering should be sufficiently large and bold to permit reduction. If requested, original drawings and photographs will be returned to the author upon publication of the paper.
7. Formulae should not be written to small and not with pencil. Separate lines for formulae are desirable. SI-units should be used.
   Letters in formulae are normally printed in italics, numbers in ordinary upright typeface. Underlining to denote special typefaces should be done in accordance with the following code:
   Italics: wavy underlined with pencil (only necessary for type written symbols in the text)
   Boldface italics (vectors): straight forward and wavy underlined
   Upright letters (all abbreviations like all units (cm, g, ...), all elements and particles (H, He, ..., n, p, ...), elementary mathematical functions like Re, Im, sin, cos, exp, ...): black underlined
   Greek letters: red underlined
   Boldface Greek letters: red interlined twice
   Upright Greek letters (symbols of elementary particles): red and black underlined
   Large letters: underlined with pencil twice
   Small letters: overlined with pencil twice
   (This will be necessary for handwritten letters that do not differ in shape, as *c C, k K, o O, p P, s S, u U, v V, w W, x X, y Y, z Z*).
   It will help the printer if position of subscripts and superscripts is marked with pencil in the following way: $a^{\uparrow}$, $b^{i}_{\downarrow}$, $M^{\uparrow}_{i\downarrow}$, $M_{\hat{i}\hat{j}}$, $W_{n}^{\hat{1}}$
   Please differentiate between following symbols: a, α; $\alpha, \propto, \infty$; α, d; $c, C, \subset$; $e, l, \mathcal{E}, \varepsilon, \in$; $k, K, \varkappa$; $\varkappa, x, X, \chi, \times$; $l, 1$; $o, O, \sigma, 0$; p, ϱ; u, U; $\cup, v, \nu, \vee$; $\varphi, \phi, \Phi, \emptyset$; $\vartheta, \theta, \Theta$.
8. Each paper should followed by a list of references with consecutive numeration. The numbers should be placed on the line between squared brackets or typewritten inclined lines. Articles in journals should be cited with author's name and abbreviated first name, title of journal volume number (underlined), year of publication (in brackets) and page number. In case of books author's name and abbreviated first name, title, place, and year of publication should be given.
9. Manuscripts should be submitted fit for printing; badly arranged manuscripts are returned.
10. Proof reading should be limited to the correction of typographical errors. Deletions and insertions are not permitted in proof reading.
11. Of every paper published the author(s) will receive 30 reprints free of charge.
12. Within the framework of legal protection the publishers have the right of publication, distribution and translation reserved. Without express permission it is not allowed to produce photocopies, microfilms etc. of this journal or parts of it.

VOLUME 33 1985 NUMBER 9
FORTSCHRITTE DER PHYSIK

Fortschr. Phys. 33 (1985) 9, 495–522

Feynman Amplitudes as Tempered Distributions

V. A. Smirnov

Nuclear Physics Institute of Moscow State University,
Moscow, USSR

Feynman amplitudes (FA) are analyzed with the help of the method which employs as much as possible the analogy between ultraviolet (UV) and infrared (IR) divergences and is based on a modified α-representation. It is proved that an analytically and/or dimensionally regularized FA is a tempered distribution of external momenta and a meromorphic function of regularizing parameters with UV and IR poles. An absolutely convergent α-representation of analytically and/or dimensionally regularized FA is derived in various analyticity domains. This representation is obtained from the α-representation in the initial domain of convergence by inserting the operator which has the same structure as the R^*-operation that is a generalization of the dimensional renormalization.

Contents

1. Introduction

In studying divergences and singularities of Feynman amplitudes (FA) various methods were applied. The most effective one is based on the α-representation (see e.g. refs. [1–17]). The purpose of this paper is to describe a modified form of α-representation

[15] which is exclusively useful when both ultraviolet (UV) and infrared (IR) divergences are involved. Furthermore, this representation is used here for analyzing the divergences of FA, including massless ones, regarded as tempered distributions. In the following paper the method will be applied to establish the convergence of renormalized FA (i.e. to prove Bogoliubov-Parasiuk theorem) in various renormalization schemes and to derive the operator product expansion in theories with massless particles. Some other applications will be discussed in the end of the present paper.

The method described employs as much as possible the analogy between UV and IR divergences of FA considered before the restriction to the mass shell (only such IR divergences will be implied). As is well known the mathematical reason for the existence of UV divergences is that the product $\prod_{l\in\Gamma} \tilde{D}_l(x_{l_i} - x_{l_f})$ corresponding to a graph Γ (here l_i and l_f are the endpoints of a line l) becomes, generally, ill-defined because the propagator in x-space

$$\tilde{D}_l(x) = \frac{im_l}{(2\pi)^2} P_l\left(-i\frac{\partial}{\partial x}\right) \frac{K_1\left(m_l\sqrt{-x^2+i0}\right)}{\sqrt{-x^2+i0}} \tag{1.1}$$

is a distribution (in eq. (1.1) P_l is a polynomial of degree a_l and K_1 is modified Bessel function of the second kind). When one integrates this product over coordinates associated with internal vertices then IR divergences appear. If FA are calculated in the momentum space IR divergences occur because the propagators

$$D_l(p) = P_l(p)\,(m_l^2 - p^2 - i0)^{-1} \tag{1.2}$$

and δ-functions taking into account the momentum conservation can not be, generally, multiplied. As to UV divergences they are exhibited due to the integration over large loop momenta.

Actually, the divergences transform into singularities in various regularizing parameters. In particular, when the analytic regularization [3] is employed and the leading singularity of x-space propagator turns to be $(x^2 - i0)^{\lambda_l - \omega_l - 1}$, UV divergences manifest themselves as poles in parameters λ_l. The existence of such poles follows from the theorem [18, ch. 3, § 4; 19] which states that the product $|Q_1(\boldsymbol{x})|^{\lambda_1} \dots |Q_n(\boldsymbol{x})|^{\lambda_n}$ with Q_i polynomials in $\boldsymbol{x} = (x_1, \dots, x_k)$ is a distribution extending meromorphically from the region $\{\boldsymbol{\lambda} \mid \operatorname{Re}\lambda_l > 0\}$. However, this result is not constructive. Furthermore, it does not take into account the integration over internal vertices and IR poles involved.

It is the α-representation that allows to establish explicitly convergence and singularity properties of FA. It has the form

$$\begin{aligned} G_\Gamma(\boldsymbol{p}, \boldsymbol{\lambda}) &= (2\pi)^4\, \delta\Big(\sum_i p_i\Big) \int_0^\infty d\boldsymbol{\alpha} \prod_l \alpha_l^{\lambda_l} f_\Gamma(\boldsymbol{p}, \boldsymbol{\alpha}) \exp\Big(-i \sum_l m_l^2 \alpha_l\Big); \\ f_\Gamma(\boldsymbol{p}, \boldsymbol{\alpha}) &= D^{-2}(\boldsymbol{\alpha})\, Z(\boldsymbol{p}, \boldsymbol{\alpha}) \exp\big(iA(\boldsymbol{p}, \boldsymbol{\alpha})/D(\boldsymbol{\alpha})\big) \end{aligned} \tag{1.3}$$

where $\boldsymbol{\alpha} = (\alpha_1, \dots, \alpha_L)$, $d\boldsymbol{\alpha} = d\alpha_1 \dots d\alpha_L$; $Z(\boldsymbol{p}, \boldsymbol{\alpha})$ is a polynomial of the external momenta $\boldsymbol{p} = (p_1, \dots, p_K)$ which includes the non-scalar factor

$$\prod_l P_l\left(-i\frac{\partial}{\partial \xi_l}\right) \exp\left\{-i(2B(p, \alpha, \xi) + K(\alpha, \xi))/D(\alpha)\right\},$$

and D, A, B, K are the standard homogeneous functions (see e.g. ref. [2]). A regularization is needed: the analytic one [3] is introduced in (1.3) by means of the parameters $\boldsymbol{\lambda} = (\lambda_1, \dots, \lambda_L)$. Different forms of the exponent in (1.3) which turn to be useful in some cases are derived in refs. [6, 8, 9].

The α-representation clearly shows the analogy between UV and IR divergences manifesting themselves as the divergences of the parametric integral at small and large α_l, respectively. Let us now describe the main ideas of the method which extends this analogy as far as possible.

The method consists of two parts. First, the integration over every parameter α_l is divided into [0, 1] and [1, ∞] (initially, the variables α_l have dimension -2 in mass units: it is understood that dimensionless α_l have already been introduced). Eventually, the integral in (1.3) may be written as a sum over the regions $\mathcal{D}(\mathcal{L}_\alpha) = \{\alpha \mid 0 \leqq \alpha_l \leqq 1, l \in \mathcal{L}_\alpha; 1 \leqq \alpha_{l'} < \infty, l' \in \mathcal{L}_\beta = \mathcal{L} \setminus \mathcal{L}_\alpha\}$, each of them being characterized by a subset of lines. Then the variables $\beta_l = 1/\alpha_l$, $l \in \mathcal{L}_\beta$ are used in $\mathcal{D}(\mathcal{L}_\alpha)$. To exhibit the divergences associated with various subgraphs a further decomposition of $\mathcal{D}(\mathcal{L}_\alpha)$ into sectors is necessary. To do this, Hepp sectors [4] $\{\alpha \mid \alpha_{\pi(1)} \leqq \alpha_{\pi(2)} \leqq \cdots\}$, denoted below as H-sectors, may be applied (here π is a permutation of lines $\mathcal{L}_\alpha$). On the other hand, one may choose the sectors $\{\alpha \mid \alpha_{\sigma(\gamma)} \leqq \alpha_{\sigma(\bar{\gamma})}, \gamma \in \mathcal{F}\}$ [3, 13] (let us call them S-sectors) associated with maximal forests $\mathcal{F}$. Here $\bar{\gamma}$ is the minimal element of $\mathcal{F}$ including the element γ and $\sigma: \mathcal{F} \to \mathcal{L}$ is the isomorphism defined as follows $\sigma(\gamma) \in \gamma$, $\sigma(\gamma) \notin \gamma'$ $\forall \gamma' \Subset \gamma$.

Continuing UV-IR analogy we introduce similarly sectors for the variables $\beta = \{\beta_l \mid l \in L_\beta\}$. These sectors may be of H- or S-type again. Any IR sector is characterized by an IR forest which consists of non-overlapping, in IR sense, IR-irreducible subgraphs. Note that IR-irreducibility is analogous to (UV) one particle-irreducibility.

Let us turn to the second part of the method. To analyze the convergence of the contribution corresponding to the region $\mathcal{D}(\mathcal{L})$ (UV case) one may use representation (1.3) because the exponent is continuous at $\alpha_l \sim 0$. Therefore, the convergence is completely governed by the factors $\mathcal{D}^{-2}$ and Z. But if the region $\mathcal{D}(\phi)$ (i.e. $L = \phi$: IR case) is considered one should take into account oscillations of the exponent in (1.3). However, it is more convenient to employ the α-representation written in x-space (for the scalar case it can be found, for instance, in ref. [6])

$$\begin{aligned} G_\Gamma(x, \lambda) &= \int d\beta \prod_l \beta_l^{-\lambda_l} \tilde{f}_\Gamma(x, \beta) \exp\Big(-i \sum_l m_l^2/\beta_l\Big); \\ \tilde{f}_\Gamma(x, \beta) &= \big(A^k(\beta)\big)^{-2} \tilde{Z}(x, \beta) \exp\big(-iU(x, \beta)/A^k(\beta)\big) \end{aligned} \tag{1.4}$$

where $\beta = (\beta_1, \ldots, \beta_L)$, $d\beta = d\beta_1 \ldots d\beta_L$; $\tilde{Z}(x, \beta)$ is a polynomial in $x = (x_1, \ldots, x_k)$ which includes an analogous non-scalar factor, and A^k, U are certain homogeneous forms [15]. Since the exponent U/A^k in (1.4) is continuous at $\beta_l \sim 0$ IR convergence is now governed by $(A^k)^{-2}$ and $\tilde{Z}$. A remarkable fact is that for any $\mathcal{D}(\mathcal{L}_\alpha)$ contributing to the integral one can always choose the mixed representation, i.e. to make the Fourier transform with respect to some momenta, in such a way that the corresponding exponent will be a continuous function of α and β [15]. Thus, the successful choice of the mixed representation allows to take into consideration all the factors in the integrand apart from the exponent. The analysis of convergence is then performed by introducing appropriate sector variables.

In particular, this method makes it possible to generalize Speer theorem [10] which states that an analytically and dimensionally regularized FA $G_\Gamma(p, \lambda)$ in the case of a graph Γ without massless tadpoles is an analytic function in some region of regularizing parameters from which it extends to a meromorphic function with UV and IR poles, the external momenta p being euclidean and non-exceptional, i.e. $p_{i_1} + p_{i_2} + \cdots \neq 0$ for every subset of indices $i_1, i_2, \ldots$ In ref. [20] a generalization of the theorem for all euclidean momenta was formulated. The technique of the present paper allows to generalize Speer's result for all external momenta: for any graph without massless tadpoles there exists a nonempty domain $\mathcal{D}_0$ of regularizing parameters where the corresponding FA is an analytic function of these and a tempered distribution of external momenta. Note that for scalar euclidean FA this conclusion was obtained in ref. [21] with the help of another technique.

The proof of the result mentioned above is based on the absolute convergence (in the weak sense) of α-representation (1.3) at $\lambda \in \mathcal{D}_0$. From this region the integral regarded as a tempered distribution extends to a meromorphic function in $\mathbf{C}^L$ with UV and IR poles.

If the parameters λ are aside from poles, but $\lambda \notin \mathcal{D}_0$, then the α-integral becomes divergent and should be understood in the sense of analytic extension. A natural problem is to derive explicitly an absolutely convergent α-representation in various regions. As for massive dimensionally regularized FA this problem was resolved by Bergère and David [12]. To obtain the desired representation it is necessary to insert into the integrand the operator

$$\mathcal{R} = \sum_{F} \prod_{\gamma \in F} (-M_\gamma) \tag{1.5}$$

which has the structure of BPHZ R-operation [1, 4, 22]. The sum in (1.5) runs over the forests of 1 PI subgraphs of the given graph, and M_γ are subtracting operators which substitute the FA G_γ by its finite Taylor series of some order in masses and external momenta. The order of the expansion, certainly, dpends on the regularizing parameters.

The method of the modified α-representation makes it possible to resolve the problem completely. Let us describe the result. The representation is obtained by inserting into the integrand the operator

$$\mathcal{R}^* = \sum_{F = F_\alpha \cup F_\beta} \prod_{\gamma \in F_\alpha} (-M_\gamma) \prod_{\gamma \in F_\beta} (-\tilde{M}_\gamma) \tag{1.6}$$

which explicitly performs the analytic extension. The sum in (1.6) goes over "generalized forests": these are of the foım $F_\alpha \cup F_\beta$ where F_α is an ordinary (i.e. UV) forest and F_β is an IR forest (see Sect. 2) including only massless subgraphs. IR subtracting operator $\tilde{M}_\gamma$ expands FA in a Taylor series in the position space, the order of the expansion depending on the regularizing parameters.

The operator in (1.6) is denoted by $\mathcal{R}^*$ because it has the same structure as R^*-operation [23] which is a generalization of the dimensional renormalization and appears to be very useful in RG-calculations, in studying the asymptotic expansion at $m \to 0$, the operator product expansion [24] etc.

The method presented here has other applications (see e.g. refs. [25] where the singularity structure of FA was analyzed). However, they are beyond the scope of this paper which deals with the problems described above. The material is organized as follows. In Section 2 the main graph-theoretical concepts are introduced. Sect. 3 contains definitions of the basic polylinear forms. Sect. 4 is devoted to a derivation of the mixed representation. The analysis of convergence and analytical properties of FA is performed in Sect. 5. In Sect. 6 the Absolutely convergent α-representation of analytically and/or dimensionally regularized FA in different regions of analyticity is derived. In Conclusion the R^*-operation is briefly discussed. Finally, in Appendices, the factorization properties of the polylinear forms are established.

2. Graph-theoretical definitions and notations

Let Γ be a connected graph with L lines $\mathcal{L}$ and V vertices $\mathcal{V}$. The graph is oriented: the corresponding incidence $V \times L$-matrix $E = \|e_{il}\|$ has elements ± 1, if the i-th vertex is the initial (final) endpoint of a line l, and $e_{il} = 0$ elsewhere.

The number of independent circuits (loops) of the graph is $\mathfrak{N} = L - V + 1$. A cut-set of a graph is a minimal set of lines such that at least one of its connectivity components becomes disconnected when these lines are removed. The graph Γ has $\mathfrak{M} = L - \mathfrak{N} = V - 1$ cut-sets [6, 26].

Let γ be a subgraph of Γ with lines $\mathcal{L}(\gamma)$. It includes all the vertices which are incident to these lines. Furthermore, in some cases (see below) isolated vertices may also belong to $\mathcal{V}(\gamma)$. Let $L(\gamma)$, $V(\gamma)$, $c(\gamma)$, $\mathfrak{N}(\gamma) = L(\gamma) - V(\gamma) + c(\gamma)$, $\mathfrak{M}(\gamma) = L(\gamma) - \mathfrak{N}(\gamma)$ be the number of lines, vertices, connectivity components, loops, and cut-sets, respectively. The subgraph $\Gamma \setminus \gamma$ composed of lines $\mathcal{L} \setminus \mathcal{L}(\gamma)$ is denoted by $\tilde{\gamma}$. The reduced graph Γ/γ is obtained from Γ by contracting every connectivity component of γ to a point. Let $\mathfrak{M}^\Gamma(\gamma)$ be the number of cut-sets of Γ which are in $\mathcal{L}(\gamma)$: the relations $\mathfrak{M}^\Gamma(\gamma) = \mathfrak{M}(\Gamma/\tilde{\gamma}) = L(\Gamma/\tilde{\gamma}) - \mathfrak{N}(\Gamma/\tilde{\gamma})$ yield the formula $\mathfrak{M}^\Gamma(\gamma) = L(\gamma) + \mathfrak{N}(\tilde{\gamma}) - \mathfrak{N} \equiv c(\Gamma/\tilde{\gamma}) - 1$.

A subgraph T is a n-tree (a tree in the case of $n = 1$) if $\mathcal{V}(T) = \mathcal{V}$, $\mathfrak{N}(T) = 0$, $c(T) = n$. For $n \geqq 2$ n-tree may involve isolated vertices. A pseudotree [6] is a subgraph with $\mathcal{V}(T) = \mathcal{V}$, $\mathfrak{N}(T) \doteq c(T) = 1$. Given two subgraphs γ_1 and γ_2 their intersection $\gamma_1 \cap \gamma_2$ is the subgraph with lines $\mathcal{L}(\gamma_1) \cap \mathcal{L}(\gamma_2)$ and vertices $\mathcal{V}(\gamma_1) \cap \mathcal{V}(\gamma_2)$. The defect of pseudo- or n-tree with respect to a subgraph γ is $\mathrm{def}_T\, \gamma = c(\gamma \cap T) - c(\gamma)$ [14, 16]. It should be stressed that the vertices of γ which are not joined by lines in $\gamma \cap T$ give contribution to $c(\gamma \cap T)$ as well.

Let $\mathcal{V}^k$ be the set of k external vertices of a given Feynman graph. The graph $\hat{\Gamma}$ [27] (also denoted by Γ^∞ [10, 13, 15, 21]) is obtained by adjoining an additional vertex $\hat{v}$ to $\mathcal{V}$ connected by one line to each vertex $v_j \in \mathcal{V}^k$, the set of these lines being denoted by $\hat{\mathcal{L}}$. Similarly the subgraphs $\hat{\gamma}$ are composed of lines $\mathcal{L}(\gamma) \cup \hat{\mathcal{L}}$. When analyzing the convergence of FA it is useful to make partial Fourier transform. Let $\mathcal{V}^r$ be the set of r external vertices regarded in the coordinate space. The graph $\hat{\Gamma}^r$ and its subgraphs $\hat{\gamma}^r$ are obtained by adjoining an additional vertex $\hat{v}^r$ connected by lines $\hat{\mathcal{L}}^r$ to the vertices $\mathcal{V}^r$. Note that $\mathcal{L}(\hat{\gamma}^r) = \mathcal{L}(\gamma) \cup \hat{\mathcal{L}}^r$. In particular, $\hat{\Gamma}^k = \hat{\Gamma}$, $\hat{\gamma}^k = \hat{\gamma}$. Let us introduce the notation $\hat{\mathfrak{M}}(\gamma) = \mathfrak{M}^{\hat{\Gamma}}(\gamma)$.

A subgraph is irreducible if it is either 2-connected or consists of a single line [10]. A subgraph is essential [14, 12] if all the external vertices belong to one of its connectivity components. The subgraph γ will be called a tadpole if it is an irreducible component of Γ, $\tilde{\gamma}$ being essential.

Let us refer to irreducible subgraphs as to UV-irreducible (UVI). For any such γ and $\gamma_{1,2} \subset \gamma$, $\mathcal{L}(\gamma_{1,2}) \neq \emptyset$, $\mathcal{L}(\gamma_1 \cap \gamma_2) = \emptyset$ the following inequality holds $\mathfrak{N}(\gamma_1) + \mathfrak{N}(\gamma_2) < \mathfrak{N}(\gamma)$. An UV-reducible subgraph decomposes naturally into UVI components: $\gamma = \bigcup\limits_i \gamma^i$, $\mathcal{L}(\gamma^i \cap \gamma^j) = \emptyset$. Note that $\mathfrak{N}(\gamma) = \sum\limits_i \mathfrak{N}(\gamma^i)$. A subgraph γ is IR-irreducible (IRI) if for any $\gamma_{1,2} \subset \gamma$, $L(\gamma_{1,2}) \neq 0$, $\mathcal{L}(\gamma_1 \cap \gamma_2) = \emptyset$ the inequality $\hat{\mathfrak{M}}(\gamma_1) + \hat{\mathfrak{M}}(\gamma_2) < \hat{\mathfrak{M}}(\gamma)$ takes place. Otherwise, it is IR-reducible and decomposes into its IRI components: $\gamma = \bigcup\limits_i \gamma^i$ with $\hat{\mathfrak{M}}(\gamma) = \sum\limits_i \hat{\mathfrak{M}}(\gamma^i)$. A subgraph γ is IRI if, and only if the reduced graph $\hat{\Gamma}/\tilde{\gamma}$ is UVI. Let us denote by $\mathcal{U}$ and $\mathcal{J}$ the sets of UVI and IRI subgraphs, respectively.

A set $F_{\alpha(\beta)}$ of UVI (IRI) subgraphs is called UV forest (IR forest) if the following conditions hold: (a) if $\gamma, \gamma' \in F_{\alpha(\beta)}$ then $\gamma \subset \gamma'$, $\gamma' \subset \gamma$ or $\mathcal{L}(\gamma \cap \gamma') = \emptyset$ (b) if $\gamma^1, \ldots, \gamma^n \in F_{\alpha(\beta)}$ satisfy $\mathcal{L}(\gamma^i \cap \gamma^j) = \emptyset$ for any $i \neq j$ then $\gamma^1 \cup \cdots \cup \gamma^n$ is UVR (IRR). Note that UV forest $\equiv$ s-family [3]. A generalized forest (or, briefly, forest) is a set $F = F_\alpha \cup F_\beta$ where $F_{\alpha(\beta)}$ is UV forest (IR forest) and $\mathcal{L}_\alpha \cap \mathcal{L}_\beta = \emptyset$ with $\mathcal{L}_{\alpha(\beta)} = \{l \mid \exists\, \gamma \in F_{\alpha(\beta)} : l \in \gamma\}$. Any maximal forest $\mathcal{F}$ is represented as $\mathcal{F}_\alpha \cup \mathcal{F}_\beta$ with $\mathcal{L}_\alpha \cup \mathcal{L}_\beta = \mathcal{L}$, $\mathcal{F}_{\alpha(\beta)}$ being UV forest (IR forest) consisting of lines $\mathcal{L}_{\alpha(\beta)}$. For given forest $\mathcal{F}$ let us define the isomorphism $\sigma : \mathcal{F} \to \mathcal{L}$ in such a way that $\sigma(\gamma)$ satisfies $\sigma(\gamma) \in \gamma \in \mathcal{F}$, $\sigma(\gamma) \notin \gamma' \in \mathcal{F}\ \forall\, \gamma' \Subset \gamma$. The inverse mapping gives the minimal element $\sigma^{-1}(l) \in \mathcal{F}$ containing the line $l \in \mathcal{L}$. Finally, the minimal element of $\mathcal{F}$ that strictly includes γ will be denoted by $\bar{\gamma}$.

3. Polylinear forms

Let $\{\beta_l \mid l \in \mathcal{L}\}$ be a set of parameters $\beta_l \geqq 0$. The $V \times V$-matrix $\mathcal{A}$ (see e.g. refs. [2, 3, 6, 15]) is defined by

$$\mathcal{A}_{ii'} = \sum_{l \in \mathcal{L}} e_{il} e_{i'l} \beta_l . \tag{3.1}$$

The property $\sum_i e_{il} = 0$ of the incidence matrix leads to $\sum_i \mathcal{A}_{ii'} = 0$. Let $\beta_{ii'}$ be the sum $\sum \beta_l$ taken over the lines which connect v_i and $v_{i'}$. We have

$$\mathcal{A}_{ii'} = \delta_{ii'} \sum_{i''} \beta_{ii''} - \beta_{ii'} . \tag{3.2}$$

Let us choose a subset $\mathcal{V}^r \subset \mathcal{V}^k$ consisting of r vertices. For convenience, the vertices $\mathcal{V}$ are numbered in the following order: $\mathcal{V}^r$, $\mathcal{V}^k \setminus \mathcal{V}^r$, and $\mathcal{V} \setminus \mathcal{V}^k$. The $(V - r) \times (V - r)$-matrix $\mathcal{A}^r$ is composed of elements (3.1) with $i, i' = r + 1, \ldots, V$. Let us now give two equivalent definitions of the basic polylinear forms A^r, $A^r_{ii'}$, $A_i^{r,j}$, $\mathcal{V}^r_{jj'}$ with $j \neq j'$; $j, j' = 1, \ldots, r$; $i, i' = r + 1, \ldots, V$.

Definition 3.1. $A^r = \det \mathcal{A}^r$; $A^r_{ii'}$ is the (i, i') cofactor of $\mathcal{A}^r$; $A_i^{r,j}$ is the determinant of the matrix obtained from $\mathcal{A}^r$ by replacing the i-th column by $(-\mathcal{A}_{r+1,j}, \ldots, -\mathcal{A}_{V,j})$, and $V^r_{jj'}$ is the determinant of the $(V - r + 1) \times (V - r + 1)$-matrix obtained from $\mathcal{A}^r$ by ascribing the row $(\mathcal{A}_{j,r+1}, \ldots, \mathcal{A}_{j,V})$ from above, and the column $(-\mathcal{A}_{j,j'}, -\mathcal{A}_{r+1,j'}, \ldots, \mathcal{A}_{V,j'})$ from the left.

Definition 3.2.

$$\begin{aligned} A^r &= \sum_{T \in T^{[r]}} \Pi_T(\boldsymbol{\beta}); & A_i^{r,j} &= \sum_{T \in T_i^{[r]j}} \Pi_T(\boldsymbol{\beta}); \\ A^r_{ii'} &= \sum_{T \in T^{[r+1]}_{ii'}} \Pi_T(\boldsymbol{\beta}); & V^r_{jj'} &= \sum_{T \in T^{[r-1]}_{jj'}} \Pi_T(\boldsymbol{\beta}) \end{aligned} \tag{3.3}$$

where

$$\Pi_T(\boldsymbol{\beta}) = \prod_{l \in T} \beta_l , \tag{3.4}$$

$\boldsymbol{\beta} = (\beta_1, \ldots, \beta_L)$; $T^{[r]}$ is the set of r-trees of Γ such that the vertices of $\mathcal{V}^r$ belong to different connectivity components; $T_i^{[r]j}$ consists of r-trees in which the vertex v_i is in the same connectivity component as $v_j \in \mathcal{V}^r$; $T^{[r+1]}_{ii'}$ is the set of $(r + 1)$-trees such that the vertices v_i and $v_{i'}$ belong to the same connectivity component, all the vertices of $\mathcal{V}^r$ being in other different connectivity components; $T^{[r-1]}_{jj'}$ consists of $(r - 1)$-trees in which the vertices $v_j, v_{j'} \in \mathcal{V}^r$ are in the same connectivity component and the other $r - 2$ vertices are in other different components.

To prove the equivalence of the definitions let us first establish it for $r = V - 1$. In this case the matrix $\mathcal{A}^{(V-1)}$ has exactly one element $\mathcal{A}^{V-1}_{VV}$ which, in accordance with (3.2), is equal to $\sum_{j=1}^{V-1} \beta_{jV}$. Thus, definition 3.1 gives

$$A^{V-1} = \sum_{j=1}^{V} \beta_{jV}, \quad A_V{}^{V-1,j} = \beta_{jV}, \quad A^{V-1}_{VV} = 1,$$

$$U^{V-1}_{jj'} = \beta_{jj'} \sum_{j''=1}^{V-1} \beta_{j''V} + \beta_{jV} \beta_{j'V} .$$

However, it may be seen that all these relations follow directly from definition 3.2 (at $r = V - 1$ r-trees consist of a single line, $(r-1)$-trees have two lines and the only $(r+1)$-tree is without lines).

Suppose now that the set $\mathcal{V}^r$ is obtained from some $\mathcal{V}^{r+1}$ by removing the $(r+1)$-st vertex. Then the following equalities hold

$$A^r_{i,r+1} = A_i^{r+1,r+1}, \qquad i = r+2, \ldots, V \tag{3.5}$$

in accordance with both definitions. Finally, the definitions lead to the relations

$$A_i^{r,j} = \sum_{i'=r+1}^{V} \beta_{i'j} A^r_{ii'}, \tag{3.6}$$

$$\sum_{j=1}^{r} A_i^{r,j} = A^r, \tag{3.7}$$

$$U^r_{jj'} = \beta_{jj'} A^r + \sum_{i=r+1}^{V} \beta_{ij} A_i^{r,j'}. \tag{3.8}$$

Here $j \neq j'$; $j, j' = 1, \ldots, r$; $i = r+1, \ldots, V$. In particular, (3.6) is the decomposition of the determinant with respect to the i-th row, and represents substitution of the summation over r-trees by the summation over $(r+1)$-trees.

Thus the polylinear forms corresponding to the set $\mathcal{V}^r$ appear to be connected by the recurrent relations with the forms which correspond to various sets $\mathcal{V}^{r+1}$ obtained from $\mathcal{V}^r$ by adjoining some vertex, these relations being the same in the case of both definitions. Therefore, the equivalence, already shown for $r = V - 1$, holds for any r.

In the next Section the polylinear forms represented according to definition 3.1 will appear. However, for the convergence analysis of the sector integrals the second definition will be more efficient.

4. Mixed representation

Let us derive the parametric representation of FA which is constructed with propagators (1.1) in the usual way, the vertices from $\mathcal{V}^r$ being regarded in the coordinate space, and the others in the momentum space. First, we use the expression for FA

$$G_\Gamma(\boldsymbol{x}, \boldsymbol{p}) = \int dx_{r+1} \ldots dx_V \exp\left[i \sum_{i=r+1}^{k} p_i x_i\right] \prod_{l \in \mathcal{L}} \frac{1}{i} D_l\left(\sum_{i=1}^{V} e_{il} x_i\right) \tag{4.1}$$

and take propagators (1.1) represented as

$$D_l(x) = \frac{1}{(8\pi)^2} P_l\left(-i \frac{\partial}{\partial \xi_l}\right) \int_0^\infty d\beta_l \exp\{-i\beta_l (x - \xi_l)^2/2 - i m_l^2/2\beta_l\}|_{\xi_l = 0}. \tag{4.2}$$

Then we interchange the order of integration over x_i and β_l, calculate the integral

$$\int dx_{r+1} \ldots dx_V \exp\left\{-\frac{i}{2} \sum_{i,i'=r+1}^{V} \mathcal{A}^r_{ii'} x_i x_{i'} + i \sum_{i=r+1}^{V} z_i x_i\right\}$$
$$= (2\pi)^{2(V-r)} (\det A^r)^{-2} \exp\left\{\frac{i\pi(V-r)}{2} + \frac{i}{2} \sum_{i,i'=r+1}^{V} (\mathcal{A}^r)^{-1}_{ii'} z_i z_{i'}\right\}$$

at $z_i = p_i - \sum_{j=1}^{r} \mathcal{A}_{ij}x_j + \sum_l e_{il}\beta_l\xi_l$ and express $\det \mathcal{A}^r$ and $(\mathcal{A}^r)^{-1}_{ii'}$ in terms of the polylinear forms. Eventually, we have the mixed representation

$$G_\Gamma(\boldsymbol{x}, \boldsymbol{p}) = \int d\beta f_\Gamma^r(\boldsymbol{x}, \boldsymbol{p}, \beta) \exp\left\{-\frac{i}{2} \sum_l m_l^2/\beta_l\right\},$$
$$f_\Gamma^r = C_\Gamma^r (A^r)^{-2} Z^r \exp\left(-\frac{i}{2} H^r\right) \tag{4.3}$$

where $C_\Gamma^r = (4\pi^2 i)^{-\mathfrak{N}(\hat{\Gamma}_r)} 2^{-L}$,

$$H^r(\boldsymbol{x}, \boldsymbol{p}, \beta) = (A^r)^{-1}\left\{\frac{1}{2}\sum_{jj'} V^r_{jj'}(x_j - x_{j'})^2 - 2\sum_{ij} A_i^{r,j} p_i x_j - \sum_{ii'} A^r_{ii'} p_i p_{i'}\right\}, \tag{4.4}$$

$$Z^r(\boldsymbol{x}, \boldsymbol{p}, \beta) = \prod_l P_l\left(-i\frac{\partial}{\partial \xi_l}\right) \exp\left\{-\frac{i}{2}(2B^r - K^r)/A^r\right\}\Bigg|_{\xi=0} \tag{4.5}$$

$$B^r = \sum_l (Q_l^r + Y_l^r)\,\xi_l, \qquad Q_l^r = \beta_l \sum_j x_j \left(\sum_{i=r+1}^{V} e_{il} A_i^{r,j} + e_{jl} A^r\right),$$
$$Y_l^r = \beta_l \sum_i p_i \sum_{i'=r+1}^{V} e_{i'l} A^r_{ii'} \tag{4.6}$$

$$K^r = \sum_{ll'} K^r_{ll'} \xi_l \xi_{l'}, \qquad K^r_{ll'} = \delta_{ll'} \beta_l A^r - \beta_l \beta_{l'} \sum_{i,i'=r+1}^{V} e_{il} e_{i'l'} A^r_{ii'}. \tag{4.7}$$

Note that if the region of the indices $j, j', \ldots$ and $i, i', \ldots$ is not specified the sum goes over $1, \ldots, r$ and $r+1, \ldots, k$, respectively. In particular, at $r = k$ the x-space representation (1.4) is obtained. At $r = 1$ with the help of the Fourier transformation with respect to x_1 and the substitution $\beta_l = \alpha_l^{-1}$ one has standard representation (1.3) where $D(\alpha) = \prod_l \alpha_l A^1(\beta)|_{\beta_l = (1/\alpha_l)}$. Note that comparing of two definitions of the form A^1 yields Kirchhof theorem: the determinant of the matrix $\mathcal{A}^1$ at $\beta_l = 1$ equals the number of the trees of the graph [28].

Integral (4.3) is a parametric representation of unregularized FA. UV (IR) divergences manifest themselves as divergences of the integral at large (small) β_l. Let us use the analytic regularization which implies inserting the factor $\prod_l \beta_l^{-\lambda_l}$ into the integrand of (4.3). For the following analysis of convergence it is efficient to split propagator (4.2) into two (IR) and (UV) integrals in accordance with the decomposition of the region $[0, \infty)$ over $[0, 1]$ and $[1, \infty)$. After that the variable $\alpha_l = 1/\beta_l$ is introduced in the second integral. Eventually, FA decomposes as the sum $\sum G_\Gamma^{\mathcal{L}_\alpha}(\boldsymbol{x}, \boldsymbol{p}; \lambda)$ over subsets of lines $\mathcal{L}_\alpha$ which are associated with α_l-integration. For definitness the lines are numbered so that the first are $\bar{l}$ lines of $\mathcal{L}_\alpha$.

The representation for $G_\Gamma^{\mathcal{L}_\alpha}(\boldsymbol{x}, \boldsymbol{p}; \lambda)$ takes the form

$$G_\Gamma^{\mathcal{L}_\alpha}(\boldsymbol{x}, \boldsymbol{p}; \lambda) = \int_0^1 d\alpha\, d\beta\, J_\Gamma^{\mathcal{L}_\alpha}(\boldsymbol{x}, \boldsymbol{p}, \alpha, \beta; \lambda) \tag{4.8}$$

where

$$J_\Gamma{}^{\mathcal{L}_\alpha}(\boldsymbol{x}, \boldsymbol{p}, \alpha, \beta; \lambda)$$

$$= \prod_{l\in\mathcal{L}_\alpha} \alpha_l{}^{\lambda_l} \prod_{l'\in\mathcal{L}_\beta} \beta_{l'}^{-\lambda_{l'}} f_\Gamma{}^r(\boldsymbol{x}, \boldsymbol{p}, \alpha, \beta) \exp\left\{-\frac{i}{2}\left(\sum_{l\in\mathcal{L}_\alpha} m_l{}^2\alpha_l + \sum_{l\in\mathcal{L}_\beta} m_l{}^2\beta_l{}^{-1}\right)\right\},$$

$$\alpha = (\alpha_1, \ldots, \alpha_{\bar{l}}), \quad \beta = (\beta_{\bar{l}+1}, \ldots, \beta_L); \; d\alpha = d\alpha_1 \ldots d\alpha_{\bar{l}}, \quad d\beta = d\beta_{\bar{l}+1} \ldots d\beta_\Gamma;$$

$$\mathcal{L}_\beta = \mathcal{L} \setminus \mathcal{L}_\alpha.$$

The function $f_\Gamma{}^r(\ldots \alpha, \beta, \ldots)$ is expressed by the same relations (4.3) to (4.7) in terms of the forms $A^r(\alpha, \beta)$, $A_i{}^{r,j}(\alpha, \beta)$, $A^r_{ii'}(\alpha, \beta)$, $U^r_{jj'}(\alpha, \beta)$ which are in turn connected with "initial" forms $A^r \equiv A^r(\beta), \ldots$ with the help of equalities of the type $A^r(\alpha, \beta) = \prod_{l'\in\mathcal{L}_\alpha} \alpha_{l'} A^r(\beta)|_{\beta_l=(1/\alpha_l), l\in\mathcal{L}_\alpha}$. All the forms depending on (α, β) at different $\mathcal{L}_\alpha$ are denoted as in the case $\mathcal{L}_\alpha = \emptyset$. However, this should not cause misunderstanding. The polylinear forms $A^r(\alpha, \beta), \ldots$ are also defined by formulae (3.3) in which it is necessary to replace $\Pi_T(\beta)$ by

$$\Pi_T(\alpha, \beta) = \prod_{l\in\mathcal{L}_\alpha\setminus T} \alpha_l \prod_{l'\in\mathcal{L}_\beta\cap T} \beta_{l'}. \tag{4.9}$$

To analyze the convergence of integral (4.8) it is useful to turn to summation over pseudo-, 1- and 2-trees of the graph $\hat{\Gamma}^r$. The following relations hold

$$A^r(\alpha, \beta) = \sum_{T\in\hat{T}^{[1]}} \Pi_T(\alpha, \beta), \qquad A_i{}^{r,j}(\alpha, \beta) = \sum_{T\in\hat{T}_i{}^{[1]j}} \Pi_T(\alpha, \beta),$$
$$A^r_{ii'}(\alpha, \beta) = \sum_{T\in\hat{T}^{[2]}_{ii'}} \Pi_T(\alpha, \beta), \qquad U^r_{jj'}(\alpha, \beta) = \sum_{T\in\hat{T}^{[0]}_{jj'}} \Pi_T(\alpha, \beta). \tag{4.10}$$

The sums go over pseudo-, 1- and 2-trees of $\hat{\Gamma}^r$ which contain all the lines $\hat{\mathcal{L}}^r$: $\hat{T}^{[1]}$ is the set of all such trees; $\hat{T}_i{}^{[1]j}$ stands for the trees in which the vertex v_i is connected with v_j by a path having no other vertices from $\mathcal{V}^r$; $\hat{T}^{[2]}_{ii'}$ is the set of 2-trees in which the vertices v_i and $v_{i'}$ are in the connectivity component without $\hat{v}^r$; $T^{[0]}_{jj'}$ represents the set of pseudo-trees containing a circuit going through v_j, $\hat{v}^r$ and $v_{j'}$.

5. Analysis of convergence

5.1. Choice of a mixed representation

To establish the conditions at which FA G_Γ is defined unambiguously as a functional on $\mathcal{S}$ it suffices to do this for every term $G_\Gamma{}^{\mathcal{L}_\alpha}$ in the sum over $\mathcal{L}_\alpha$. Let us regard integral (4.8) as a parametric representation of a tempered distribution. In other words, let us try to define the action of the functional $G_\Gamma{}^{\mathcal{L}_\alpha}(\boldsymbol{x}, \boldsymbol{p}, \lambda)$ on a test function $\varphi \in \mathcal{S}(R^{4k})$ according to the rule

$$\langle G_\Gamma{}^{\mathcal{L}_\alpha}, \varphi\rangle = \int d\alpha\, d\beta \int d\boldsymbol{x}\, d\boldsymbol{p}\, J_\Gamma{}^{\mathcal{L}_\alpha}\varphi(\boldsymbol{x}, \boldsymbol{p}). \tag{5.1}$$

Since the Fourier transform, certainly, exists for every functional from $\mathcal{S}'$ it does not matter whether to define a tempered distribution or its Fourier transform with respect to some variables. We use this freedom to simplify as much as possible the convergence analysis of the parametric representation. For given $\mathcal{L}_\alpha$ let us choose the set $\mathcal{V}^r$ (i.e. turn to variables $\boldsymbol{x}$ and $\boldsymbol{p}$) so that the function $H^r(\boldsymbol{x}, \boldsymbol{p}, \alpha, \beta)$ in the exponent would be bounded. As a result the convergence of the integral would be governed by another part of the integrand.

In particular, for the case $\mathcal{L}_\alpha = \mathcal{L}$ one may choose "momentum" representation (1.3) or any representation in which exactly one of the external vertices is regarded in the coordinate space: the structure of the integrand is the same in all these cases. Then, to show that the exponent is bounded and the integral is convergent at small α_l (UV convergence), one should use the standard method of integration over sectors of H- or S-type. To choose an appropriate mixed representation and to analyze the convergence let us decompose the integration region into H-sectors

$$\Delta_{\pi,\bar{l}} = \left\{(\alpha, \beta) \mid \alpha_{\pi(1)} \leqq \cdots \leqq \alpha_{\pi(\bar{l})} \leqq 1 \leqq \frac{1}{\beta_{\pi(\bar{l}+1)}} \leqq \cdots \leqq \frac{1}{\beta_{\pi(L)}}\right\},$$

each of them corresponding to a permutation π and a number $\bar{l}$. Without loss of generality consider only the sector

$$\Delta = \left\{(\alpha, \beta) \mid \alpha_1 \leqq \cdots \leqq \alpha_{\bar{l}} \leqq 1 \leqq \frac{1}{\beta_{\bar{l}+1}} \leqq \cdots \leqq \frac{1}{\beta_L}\right\}. \tag{5.2}$$

Now we introduce H-sector variables (H-variables) $\boldsymbol{t} = (t_1, \ldots, t_{\bar{l}})$ and $\boldsymbol{\tau} = (\tau_{\bar{l}+1}, \ldots, \tau_L)$:

$$\begin{aligned} t_l &= \alpha_l/\alpha_{l+1}; \qquad l = 1, \ldots, \bar{l} - 1; \qquad t_{\bar{l}} = \alpha_{\bar{l}}; \\ \tau_l &= \beta_l/\beta_{l-1}; \qquad l = L, L-1, \ldots, \bar{l} + 2; \qquad \tau_{\bar{l}+1} = \beta_{\bar{l}+1} \end{aligned} \tag{5.3}$$

which belong to the segment [0, 1].

By this procedure it is possible to exhibit the behaviour of the forms at $\alpha_l, \beta_{l'} \sim 0$ and, eventually, to reduce the convergence analysis to the power counting in one-dimensional integrals. Let us use the factorization in H-variables which is proved in Appendix A. Representations (A.4)—(A.7) lead to the following factorization

$$\begin{aligned} &H^r(\boldsymbol{x}, \boldsymbol{p}, \boldsymbol{t}, \boldsymbol{\tau}) \\ &= (1 + P^r)^{-1} \left\{\frac{1}{2} \sum_{jj'} (x_j - x_{j'})^2 P^{r,jj'} \prod_l t_l^{d^{r,jj'}(\gamma_l-1)} \prod_{l'} \tau_{l'}^{d^{r,jj'}(\gamma_{l'-1})} \right. \\ &\quad \left. - 2 \sum_{ij} x_j p_i P_i^{r,j} \prod_l t_l^{d_i^{r,j}(\gamma_l)} \prod_{l'} \tau_{l'}^{d_i^{r,j}(\gamma_{l'-1})} - \sum_{ii'} p_i p_{i'} P_{ii'}^r \prod_l t_l^{d_{ii'}^r(\gamma_l)} \prod_{l'} \tau_{l'}^{d_{ii'}^r(\gamma_{l'-1})-1} \right\}. \end{aligned} \tag{5.4}$$

Here γ_l ($l = 1, \ldots, L$) are the subgraphs composed of lines $\{1, \ldots, l\}$. The definitions of $d^{r,jj'}$, $d_i^{r,j}$, $d_{ii'}^r$ are given in Appendix A. If a region of indices l and l' is not specified they take the values $1, \ldots, \bar{l}$ and $\bar{l} + 1, \ldots, L$, respectively. As may be seen from (5.4) the function H^r will be continuous if the following inequalities hold

$$d^{r,jj'}(\gamma_l) \geqq 1, \qquad d_{ii'}^r(\gamma_{l'-1}) \geqq 1. \tag{5.5}$$

Let us now describe the proper choice of the set $\mathcal{V}^r$ corresponding to given $\mathcal{L}_\alpha$. Include into $\mathcal{V}^r$ a vertex from every set of external vertices which are connected with each other by the paths composed of lines $\mathcal{L}_\alpha$. As a result conditions (5.5), as well as analogous conditions in the case of other sectors, will be fulfilled. Indeed, any subgraph $\gamma_l \subset \Gamma_\alpha$ (Γ_α consists of lines $\mathcal{L}_\alpha$ and all the vertices of Γ) has the vertices $\mathcal{V}^r$ in different connectivity components: hence, the first relations in (5.5) are valid. As to the second relations, they hold because any external vertex $v_i \notin \mathcal{V}^r$ is connected in Γ_α with some vertex from $\mathcal{V}^r$ and, therefore, in any subgraph $\gamma_{l'-1}$.

5.2. Power counting of sector variables

Thus, the proper choice of mixed representation (4.8) simplifies the convergence analysis of $G_{\Gamma}{}^{\mathcal{L}_\alpha}$. In fact, the convergence is governed by the massive exponent and other factors which consist of

1) $\prod_l t_l^{\nu_l} \prod_{l'} \tau_{l'}^{-\bar{\nu}_{l'}}$ with $\nu_l = \sum_{l''=1}^{l} \lambda_{l''}$, $\bar{\nu}_{l'} = \sum_{l''=l'}^{L} \lambda_{l''}$ 2) $\prod_l t_l{}^{l-1} \prod_{l'} \tau_{l'}^{L-l'}$ — the Jakobian corresponding to sector variables $(\boldsymbol{t}, \boldsymbol{\tau})$; 3) $\left(A^r(\alpha, \beta)\right)^{-2}$; 4) $B^r_{l_0}/A^r$ and $K^r_{l_1 l_2}/A^r$ which appear due to differentiation with respect to $\boldsymbol{\xi}$. It is easy to notice that $\mathfrak{N}(\hat{p}_l{}^r) = \mathfrak{N}(\gamma_l)$, $\mathfrak{M}^{\hat{\Gamma}^r}(\gamma'_{l'}) = \hat{\mathfrak{M}}(\gamma'_{l'})$ because of the choice of the representation. Here $\gamma'_{l'}$ are the subgraphs composed of lines $\{l', \ldots, L\}$. Taking into account factorization (A.4) of the form A^r one has that the first three of the factors involved become

$$\prod_l t_l{}^{\nu_l - 2\mathfrak{N}(\gamma_l) + l - 1} \prod_{l'} \tau_{l'}^{-\bar{\nu}_{l'} - 2\hat{\mathfrak{M}}(\gamma'_{l'}) + L - l'} (1 + P_A)^{-2}. \tag{5.6}$$

Now we consider the factors contributing to Z^r and represented according to (A.8) to (A.10). Note that the proper choice of the mixed representation leads to the following relations

$$\min \{\psi^r_{l_0}(\gamma_l), y^r_{l_0}(\gamma_l)\} = y^1_{l_0}(\gamma_l),$$

$$\min \{1 - \psi^r_{l_0}(\gamma_{l'-1}), y^r_{l_0}(\gamma_{l'-1})\} = 1 - \psi^k_{l_0}(\gamma_{l'-1}),$$

$$\chi^r_{l_1 l_2}(\gamma_l) = \chi^1_{l_1 l_2}(\gamma_l), \qquad \chi^r_{l_1 l_2}(\gamma_{l'-1}) \leqq \chi^k_{l_1 l_2}(\gamma_{l'-1}).$$

Therefore, the terms presenting in the exponent and depending on $\boldsymbol{\xi}$ take the form

$$\sum_{l_0} \xi_{l_0} \prod_l t_l{}^{y^1_{l_0}(\gamma_l)} \prod_{l'} \tau_{l'}^{1 - \psi^k_{l_0}(\gamma_{l'-1})} A^r_{l_0}(\boldsymbol{x}, \boldsymbol{p}, \boldsymbol{t}, \boldsymbol{\tau})$$
$$+ \sum_{l_1 l_2} \xi_{l_1} \xi_{l_2} \prod_l t_l{}^{-\chi^1_{l_1 l_2}(\gamma_l)} \prod_{l'} \tau_{l'}^{1 - \chi^k_{l_1 l_2}(\gamma_{l'-1})} A^r_{l_1 l_2}(\boldsymbol{t}, \boldsymbol{\tau})$$

where $A^r_{l_0}$ and $A^r_{l_1 l_2}$ are smooth functions. After differentiating in $\boldsymbol{\xi}$ at $\boldsymbol{\xi} = 0$ the sum of various terms appears in the integrand. To achieve conditions of convergence it is necessary to choose the term containing the minimal degrees of the variables t_l and $\tau_{l'}$.

Let us introduce additional notation. The set of UVI components of a subgraph γ which are 1 PI and do not include the vertex $\hat{v}$ will be denoted by $\hat{\mathscr{K}}(\gamma)$, and $\hat{w}(\gamma)$ will be the union of the rest 1 PI components. Let $\hat{\mathscr{B}}(\gamma)$ be the set of 1 PR UV-components of $\hat{p}$. Furthermore,

$\mathrm{Odd}(w) = \sum_{l \in w} a_l - 2\left[\sum_{l \in w} a_l/2\right]$ will stand for the oddness of UV-irreducible component w of γ and the square brackets will always mean the integer part of a number.

For simplicity we suppose that polynomials $P_l(p)$ in propagators are monomials $(p_\mu)^{a_l}$. Let us now describe how one should act on the exponent by the operator $X = \prod_{l \in \mathcal{L}} \left(-i(\partial/\partial \xi_l)\right)^{a_l}$ to obtain in the integrand the factor which includes the minimal power of a variable $\tau_{l'}$. Pick out the product over $\hat{w}(\gamma_{l'-1})$ and act on the linear terms of the exponent: the power of $\tau_{l'}$ in the integrand will not be less than before this procedure. Let us represent the rest part of X as $X_1 X_2$ in accordance with decomposition a_l into the sum $a_l = 2\,[a_l/2] + \mathrm{Odd}\,(l)$. Act by X_1 on diagonal terms (quadratic in $\boldsymbol{\xi}$)

of the exponent. The power of $\tau_{l'}$ will be increased by

$$\sum_{l\in\gamma'_{l'}\cup\hat{\mathscr{B}}(\gamma_{l'-1})} [a_l/2]. \tag{5.7}$$

By means of X_2 let us differentiate quadratic terms corresponding to lines l_1 and l_2 which belong to the same $w \in \hat{\mathscr{K}}(\gamma_{l'-1})$: the power of $\tau_{l'}$ will not be changed. After that one should differentiate by the rest factors

$$\prod_{l\in\gamma'_{l'}\cup\mathscr{B}(\gamma_{l'-1})} \left(-i\frac{\partial}{\partial\xi_l}\right)^{\mathrm{Odd}(l)} \prod_{w\in\hat{\mathscr{K}}(\gamma_{l'-1})} \left(-i\frac{\partial}{\partial\xi_{l_w}}\right)^{\mathrm{Odd}(w)}$$

(here l_w is a line of w) the quadratic terms of the exponent. Finally, the linear term is differentiated exactly one time if the degree corresponding to lines $\mathscr{L}\setminus\mathscr{L}(\hat{w}(\gamma_{l'-1}))$ is odd. As a result the power of $\tau_{l'}$ is increased by

$$\left[\frac{1}{2}\left(\sum_{l\in\gamma'_{l'}\cup\hat{\mathscr{B}}(\gamma_{l'-1})} \mathrm{Odd}\,(l) + \sum_{w\in\hat{\mathscr{K}}(\gamma_{l'-1})} \mathrm{Odd}\,(w) + 1\right)\right]. \tag{5.8}$$

One may verify that such way of differentiation produces the term which has the minimal powers of $\tau_{l'}$. To do this, it is necessary to consider separately the action of X_1 and X_2: in the first case the contribution to the power of $\tau_{l'}$ is not less than (5.7) and, in the second case, the contribution is not less than (5.8). Combining these inequalities we see that the variable $\tau_{l'}$ enters with the power $-\bar{\nu}_{l'} - \Omega^\Gamma(\gamma'_{l'}) - 1$ where

$$\Omega^\Gamma(\gamma) = 2\mathfrak{M}(\gamma) - L(\gamma) - \left[\frac{1}{2}\left(\sum_{l\in\gamma\cup\hat{\mathscr{B}}(\bar{\gamma})} a_l + \sum_{w\in\hat{\mathscr{K}}(\bar{\gamma})} \mathrm{Odd}\,(w) + 1\right)\right]. \tag{5.9}$$

Note that the scalar part of $\Omega^\Gamma(\gamma)$ may be also written as $L(\gamma) - 2\mathfrak{N}(\hat{\Gamma}/\bar{\gamma})$. The following inequality holds $\Omega^\Gamma(\gamma) \leqq [(\tilde{\omega}(\gamma)+1)/2]$ where

$$\tilde{\omega}(\gamma) = -\omega(\hat{\Gamma}/\bar{\gamma}) \tag{5.10}$$

is IR-index of divergence.

The power counting of the variables t_l is performed analogously. For simplicity we do not take into account differentiation of the linear term of the exponent: this does not diminish the powers. Thus, to exhibit the minimal power of t_l it is necessary first to differentiate quadratic (in ξ) terms which correspond to lines l_1 and l_2 belonging to the same 1 PI UVI component of subgraph γ_l. A straightforward calculation shows that the power of t_l is $\nu_l - \overline{\omega}(\gamma_l) - 1$ with

$$\overline{\omega}(\gamma) = 2\mathfrak{N}(\gamma) - L(\gamma) + \sum_{w\in\mathscr{K}(\gamma)}\left[\frac{1}{2}\sum_{l\in w} a_l\right] \leqq \left[\frac{\omega(\gamma)}{2}\right] \tag{5.11}$$

and $\mathscr{K}(\gamma)$ stands for the set of 1 PI UVI components of γ.

Formulae (5.9) and (5.11) are easily generalized to the case of arbitrary polynomials P_l. The expression for $\overline{\omega}$ will include the maximal powers of P_l and the minimal powers will contribute to Ω^Γ. Furthermore, if propagators of the type $D_l = P_l(p)/(m_l^2 - p^2 - i0)^{1+n_l}$ are concerned one should add $\pm\sum_{l\in\gamma} n_l$ to the expression for $\Omega(\overline{\omega})$.

5.3. Speer theorem and conditions of convergence

Thus, the contribution of sector (5.2) to $G_\Gamma^{\mathcal{L}_\alpha}$ looks like

$$G_\Gamma^{\mathcal{L}_\alpha;\Delta}(\boldsymbol{x},\boldsymbol{p};\lambda) = \int_0^1 d\alpha\, d\beta \prod_l t_l^{\nu_l-\overline{\omega}(\gamma_l)-1} \times \prod_{l'} \tau_{l'}^{-\tilde{\nu}_{l'}-\Omega^\Gamma(\gamma'_{l'})-1} Q(\boldsymbol{x},\boldsymbol{p},\boldsymbol{t},\boldsymbol{\tau}) \exp\left\{-i\sum_{l\in\mathcal{L}} m_l^2\alpha_l(\boldsymbol{t},\boldsymbol{\tau})\right\} \tag{5.12}$$

with $Q \in C^\infty$. This integral is absolutely convergent if

$$\operatorname{Re}\nu_l > \overline{\omega}(\gamma_l), \qquad \operatorname{Re}\tilde{\nu}_{l'} < -\Omega^\Gamma(\gamma'_{l'}). \tag{5.13}$$

It should be noted that (5.12), as well as any other sector contribution, turns to be infinitely differentiable function with respect to chosen variables $(\boldsymbol{x},\boldsymbol{p})$. Obviously, the derivatives of $G_\Gamma^{\mathcal{L}_\alpha}$ in $\boldsymbol{x}$ and $\boldsymbol{p}$ are polynomially bounded. Therefore, integral (5.12) under conditions (5.13) defines a tempered distribution. These conditions give also the analyticity region in λ.

Taking into account the decomposition associated with various sets $\mathcal{L}_\alpha$ and H-sectors one obtains the restictions on analyticity region $\mathcal{D}_\Gamma$ of regularized FA regarded as a tempered distribution

$$\overline{\omega}(\gamma) < \operatorname{Re}\nu(\gamma) < -\Omega^\Gamma(\gamma), \qquad \gamma \subset \Gamma. \tag{5.14}$$

Here $\nu(\gamma) = \sum_{l\in\gamma}\lambda_l$. An analysis of sector integrals involved shows that the regularized FA extends from (5.14) to a meromorphic function in $\mathbf{C}^L$ with UV and IR poles

$$\nu(\gamma) = \overline{\omega}(\gamma) - n, \qquad n = 0, 1, \dots \tag{5.15}$$

$$\nu(\gamma) = -\Omega^\Gamma(\gamma) + n, \qquad n = 0, 1, \dots \tag{5.16}$$

If there are non-zero masses some IR poles dissappear. In fact, a mass $m_l \neq 0$ allows to extend (5.12) to $\tilde{\nu}_{l'}$ with $\operatorname{Re}\tilde{\nu}_{l'} \gg 0$. To do this, one may use integration by parts [15] in sector variables. Therefore, conditions (5.16) and rhs of (5.14) should involve only massless subgraphs.

Let us prove that for graphs without massless tadpoles the region (5.14) is not empty. First, define (see Speer [10]) for given graph

$$\lambda_l^{(0)\Gamma} = 1 + 2\delta - 2\Big(\sum_{T\in T_l^{[1]}} 1\Big)\Big/\Big(\sum_{T\in T^{[1]}} 1\Big) \tag{5.17}$$

with $T_l^{[1]}$ the set of trees including a line l. Then we will show that for sufficiently small $\delta > 0$ parameters (5.17) satisfy

$$2\mathfrak{N}(\gamma) - L(y) < \sum_{l\in\gamma}\lambda_l^{(0)\Gamma} < 2\big(\mathfrak{N}(\hat{\Gamma}) - \mathfrak{N}(\hat{\bar{\gamma}})\big) - L(\gamma). \tag{5.18}$$

The left part of the inequalities is established in [10] for any $\delta > 0$. As for the right part, we insert $\lambda_l^{(0)\Gamma}$ into (5.18) and use the relation

$$\sum_{T\in T^{[1]}} L(\gamma\cap T) = L - \mathfrak{N} - \sum_{T\in T^{[1]}} L(\bar{\gamma}\cap T).$$

As a result we have

$$\sum_{T\in T^{[1]}} L(\tilde\gamma \cap T)\Big/\Big(\sum_{T\in T^{[1]}} 1\Big) < \mathfrak{N}(\hat\Gamma) - \mathfrak{N}(\Gamma) - \mathfrak{N}(\hat{\tilde\gamma}) + L(\tilde\gamma) - \delta L(\gamma). \tag{5.19}$$

Formula (A.2) gives $L(\tilde\gamma \cap T) = L(\tilde\gamma) - \mathfrak{N}(\tilde\gamma) - \mathrm{def}_T\,(\tilde\gamma)$ whence

$$\sum_{T\in T^{[1]}} L(\tilde\gamma \cap T)\Big/\Big(\sum_{T\in T^{[1]}} 1\Big) \leqq L(\tilde\gamma) - \mathfrak{N}(\tilde\gamma). \tag{5.20}$$

The equality holds if $\mathrm{def}_T\,\tilde\gamma = 0$ for any $T \in T^{[1]}$. Such situation is possible only when $\tilde\gamma$, as well as γ, is a union of UVI components of Γ. If this was not the case then there would be a circuit including lines both from $\mathscr{L}(\gamma)$ and $\mathscr{L}(\tilde\gamma)$. Therefore, by the standard procedure (see e.g. the proof of the factorization of A^r in Appendix A) one would construct a tree such that $\mathrm{def}_{\tilde\gamma}\, T \neq 0$.

Thus, to establish (5.19) it is sufficient to show that

$$\mathfrak{N}(\hat{\tilde\gamma}) - \mathfrak{N}(\tilde\gamma) \leqq \mathfrak{N}(\hat\Gamma) - \mathfrak{N}(\Gamma) - \delta L(\gamma). \tag{5.21}$$

If the term with δ is ignored the equality may hold only for essential $\tilde\gamma$. Hence, (5.20) and (5.21) take place, provided γ is a massless tadpole. Therefore, for sufficiently small $\delta > 0$ strict inequalities (5.19) equivalent to rhs inequalities (5.18) are valid. Define then

$$\lambda_l{}^\Gamma = \lambda_l{}^{(0)\Gamma} + a_l/2. \tag{5.22}$$

Inequalities (5.14) for these parameters follow directly from (5.18). Thus, $\lambda^\Gamma \in \mathscr{D}_\Gamma$.

Let us now describe the method in the case of S-sectors

$$\mathscr{D}(\mathscr{F}) = \{(\alpha, \beta) \mid \alpha_l \leqq \alpha_{\sigma(\gamma)} \leqq 1;\, l \in \gamma \in \mathscr{F}_\alpha;\, \beta_l \leqq \beta_{\sigma(\gamma)} \leqq 1,\, l \in \gamma \in \mathscr{F}_\beta,$$
$$\mathscr{F} = \mathscr{F}_\alpha \cup \mathscr{F}_\beta\}.$$

Any S-sector is associated with a maximal forest $\mathscr{F}$. The union of all the sectors equals the whole integration domain and the intersection of any pair of the sectors has zero measure. For given $\mathscr{D}(\mathscr{F})$ we introduce S-sector variables (S-variables) $\boldsymbol{t}' = \{t_\gamma' \mid \gamma \in \mathscr{F}_\alpha\}$ and $\boldsymbol{\tau}' = \{\tau_\gamma' \mid \gamma \in \mathscr{F}_\beta\}$ as follows

$$\alpha_l = \prod_{l\in\gamma\in\mathscr{F}_\alpha} t_\gamma',\ l \in \mathscr{L}_\alpha; \qquad \beta_l = \prod_{l\in\gamma\in\mathscr{F}_\beta} \tau_\gamma',\ l \in \mathscr{L}_\beta. \tag{5.23}$$

Conversely, $t_\gamma' = \alpha_{\sigma(\gamma)}/\alpha_{\sigma(\tilde\gamma)}$ or $\tau_\gamma' = \beta_{\sigma(\gamma)}/\beta_{\sigma(\tilde\gamma)}$ if γ is non-maximal, and $t_\gamma' = \alpha_{\sigma(\gamma)}$ or $\tau_\gamma' = \beta_{\sigma(\gamma)}$ elsewhere. The Jakobian looks like

$$\prod_{\gamma\in\mathscr{F}_\alpha} (t_\gamma')^{L(\gamma)-1} \prod_{\gamma\in\mathscr{F}_\beta} (\tau_\gamma')^{L(\gamma)-1}.$$

The factorization of A^r in S-variables is proved in Appendix B. As to other forms, the factorization is easily generalized to this case. Therefore, the function f^r from (4.8) becomes

$$f^r(\boldsymbol{x}, \boldsymbol{p}, \boldsymbol{t}', \boldsymbol{\tau}') = \prod_{\gamma\in\mathscr{F}_\alpha} (t_\gamma')^{-[\omega(\gamma)/2]-L(\gamma)} \prod_{\gamma'\in\mathscr{F}_\beta} (\tau_{\gamma'}')^{-\Omega^\Gamma(\gamma')-L(\gamma')}\, Q(\boldsymbol{x}, \boldsymbol{p}, \boldsymbol{t}', \boldsymbol{\tau}') \tag{5.24}$$

with $Q \in C^\infty$.

It should be emphasized that in the following Section, and in the following paper, $(\boldsymbol{t}', \boldsymbol{\tau}')$ will be auxiliary rather than integration variables. Representation (5.24) will play a desive role when proving the basic results.

Note that by using S-sectors instead of H-sectors one comes to more exact restrictions on the pole structure of FA. In fact, relations (5.15) and lhs of inequalities (5.14) should be written only for $\gamma \in \mathcal{U}$. Similarly, only $\gamma \in \mathcal{I}$ should enter into (5.16) and rhs of (5.14). Ler us denote by $\mathcal{D}_0$ the corresponding reduction of domain (5.14).

S-sectors may be also associated with the forests with modified definition of their UV part $\mathcal{F}_\alpha$. It is possible to regard a subgraph as UVI if it is 1 PI or consists of a single line. The initial definition is natural from the mathematical point of view while the second definition is of use in studying the R-operation [13]. One can easily verify that representation (5.24) holds in this case as well.

Let us now formulate the theorem which involves the results of this Section and is a generalization of Speer theorem [10] to the case of all external momenta.

Theorem 5.1. The analytically regularized FA $G_\Gamma(\boldsymbol{p}, \lambda)$ corresponding to a graph without massless tadpoles is a tempered distribution of $\boldsymbol{p}$ and an analytic function of λ in the domain $\mathcal{D}_0$ from which it extends meromorphically with UV and IR poles.

Corollary. If $\overline{\omega}(\gamma) < 0$ for any $\gamma \in U$ then UV divergences are absent. If $\Omega^\Gamma(\gamma) < 0$ for any massless $\gamma \in \mathcal{I}$ then IR divergences are absent.

The numbers $\overline{\omega}$ and Ω may be regarded as modified indices of divergences. Of course, the negativity of indices ω and $\tilde{\omega}$ is also sufficient for convergence.

It should be noted that usually FA is considered as divergent if it can not be unambiguously defined at non-exceptional euclidean momenta. Then one can not also define FA as a functional on a subspace of test functions which vanish in a vicinity of exceptional momenta [16]. All additional ("superficial") IR divergences manifest themselves if FA is integrated with smooth functions. In particular, when FA is calculated in the coordinate space (integration with $e^{i\boldsymbol{px}}$) both usual and additional IR divergences are equally exhibited.

The inequalities $\Omega^\Gamma(\gamma) < 0$ written at essential $\tilde{\gamma}$ provide more strict IR-convergence conditions (in the usual sense) than in refs. [16, 29]. Nevertheless, the conditions of ref. [16] turned to be sufficient to classify theories from the point of view of IR divergences.

6. Absolutely convergent α-representation of analytically and/or dimensionally regularized Feynman amplitudes

6.1. Review

As was shown in the previous Section representation (1.3) is absolutely convergent at $\lambda \in \mathcal{D}_0$. In this Section a representation which is absolutely convergent almost in the whole analyticity domain will be derived. In other words, the analytic continuation procedure of α-integral (1.3) regarded as a tempered distribution will be explicitly performed.

Suppose that at some $\lambda \notin \mathcal{D}_0$ integral (1.3) is not absolutely convergent and for any $\gamma \in U$ ($\gamma \in \mathcal{I}$) the following inequalities hold

$$\operatorname{Re} \nu(\gamma) \neq [\omega(\gamma)/2] - n \qquad (\neq -\Omega^\Gamma(\gamma) + n), \qquad n = 0, 1, \dots$$

Then there exist numbers $\boldsymbol{n} = \{n_\gamma \mid \gamma \in U\}$ and $\boldsymbol{n} = \{\tilde{n}_\gamma \mid \gamma \in \mathcal{I}\}$ such that

$$n_\gamma < [\omega(\gamma)/2] - \operatorname{Re} \nu(\gamma) < n_\gamma + 1,\ \tilde{n}_\gamma < \Omega^\Gamma(\gamma) + \operatorname{Re} \nu(\gamma) < \tilde{n}_\gamma + 1. \tag{6.1}$$

Let us denote by $\mathcal{D}(\boldsymbol{n}, \boldsymbol{n})$ the set of parameters λ which fulfil (6.1). Notice that $\Omega^\Gamma(\gamma) \leqq [(\tilde{\omega}(\gamma) + 1)/2]$, but, in this Section, it suffices to imply the sign of equality.

Now we define the operator M_γ^a which transforms FA G_γ into the sum of the first a terms of Taylor expansion in external momenta and internal masses. In the language of α-representation M_γ^a acts as follows [2]

$$M_\gamma^a G_\Gamma = (2\pi)^4 \,\delta\Big(\sum_i p_i\Big) \int d\alpha \prod_l \alpha_l^{\lambda_l} \mathcal{M}_{\varkappa_\gamma}^{a+R} \varkappa_\gamma^{4\mathfrak{N}(\gamma)+a(\gamma)+R} \times \Big\{f_\Gamma(\boldsymbol{p}, \boldsymbol{\alpha}) \exp\Big(-i \sum_{l\in\mathcal{L}} m_l^2 \alpha_l\Big)\Big\}\Big|_{\alpha_l \to \varkappa_\gamma^2 \alpha_l, l\in\gamma}.$$

Here $a(\gamma) = \sum_{l\in\gamma} a_l$, R is a sufficiently large number and $\mathcal{M}_\varkappa^a g(\varkappa) = \sum_{0\le k\le a} g^{(k)}(0)/k!$ Define then, similarly, the operator $\tilde{M}_\gamma^{\tilde{a}}$ which perfoıms Taylor expansion in the coordinate space. For convenience we use "β-parametric" representation (1.4) (i.e. $\alpha_l \to \beta_l = 1/\alpha_l$)

$$\tilde{M}_\gamma^{\tilde{a}} G_\Gamma = (2\pi)^4 \,\delta\Big(\sum_i p_i\Big) \int d\beta \prod_l \beta_l^{-\lambda_l} \mathcal{M}_{\tilde{\varkappa}_\gamma}^{\tilde{a}+R} \tilde{\varkappa}_\gamma^{4\mathfrak{M}(\gamma)-a(\gamma)+R} \times f_\Gamma(\boldsymbol{p}, \boldsymbol{\beta})\big|_{\beta_l \to \tilde{\varkappa}_\gamma^2 \beta_l, l\in\gamma} \exp\Big(-i \sum_{l\in\mathcal{L}} m_l^2 \beta_l^{-1}\Big).$$

Let us now state the main result of this Section for massless FA.

Theorem 6.1. Analytic extension of FA (1.3) regarded as a tempered distribution is given at $\boldsymbol{\lambda} \in \mathcal{D}(\boldsymbol{n}, \tilde{\boldsymbol{n}})$ by the absolutely convergent integral

$$G_\Gamma(\boldsymbol{p}, \boldsymbol{\lambda}) = (2\pi)^4 \,\delta\Big(\sum_i p_i\Big) \int d\alpha \prod_{l\in\mathcal{L}} \alpha_l^{\lambda_l} \,\mathcal{R}^* f_\Gamma(\boldsymbol{p}, \boldsymbol{\alpha}). \tag{6.2}$$

The operator $\mathcal{R}^*$ looks like

$$\mathcal{R}^* = \sum_{F=F_\alpha \cup F_\beta} \prod_{\gamma\in F_\alpha} (-M_\gamma^{a^\gamma}) \prod_{\gamma\in F_\beta} (-\tilde{M}_\gamma^{\tilde{a}^\gamma}) \tag{6.3}$$

with $a^\gamma = \omega(\gamma) + 2[-\mathrm{Re}\, \nu(\gamma)]$, $\tilde{a}^\gamma = \tilde{\omega}(\gamma) + 1 + 2[\mathrm{Re}\, \nu(\gamma)]$, $\overset{(\sim)}{n}_\gamma = [\overset{(\sim)}{a}{}^\gamma/2]$. The sum in (6.3) runs over the forests of a given graph.

If there are massive lines then the condition of the theorem changes as follows. The sum should go over the forests F with IR part F_β involving only massless subgraphs. Then the assertion of the theorem holds both for euclidean FA and for FA with IR ε – regularization in Minkowski space. This statement follows directly from the convergence analysis and is not discussed below.

A proof of the theorem will be given in Subsect. 6.2 and 6.3. First, the absolute convergence of integrals (6.2) will be established. Then it will be shown that all these integrals may be obtained from each other by analytic continuation procedure.

6.2. Absolute convergence

As in Sect. 5 let us decompose the integration region in (6.2) into H-sectors

$$\Delta_{\pi,\bar{l}} = \{\alpha \mid \alpha_{\pi(1)} \leqq \cdots \leqq \alpha_{\pi(\bar{l})} \leqq 1 \leqq \alpha_{\pi(\bar{l}+1)} \leqq \cdots \leqq \alpha_{\pi(L)}\}, \tag{6.4}$$

each of them being characterized by a permutation π of the numbers $1, \ldots, L$ and a number $\bar{l} = 0, 1, \ldots, L$. When proving the convergence consider only the sector $\Delta \equiv \Delta_{\bar{l}}$ associated with identical permutation. Let $\mathcal{N}$ be the nest composed of the subgraphs γ_l, $l = 1, \ldots, L$. For $l \geqq \bar{l} + 1$ introduce the variables $\beta_l = 1/\alpha_l$.

The contribution of the sector Δ to (6.2) will be denoted by G_Δ. When analyzing convergence of various contributions to G_Δ it is very convenient to use mixed representation

(4.8). However, the choice of the latter will be determined both by the number $\bar{l}$ and by a contribution of the terms from (6.3). It should be emphasized that representation (4.8) will correspond to a set $\mathscr{L}_\alpha$ which, generally, will not coincide with $\{1, \ldots, l\}$ (see below).

Since the parameters $\varkappa = \{\varkappa_\gamma \mid \gamma \in F_\alpha\}$ and $\tilde{\varkappa} = \{\tilde{\varkappa}_\gamma \mid \gamma \in F_\beta\}$ associated with operators M_γ and $\tilde{M}_\gamma$ are introduced than the functions A^τ, Z^τ, H^τ from (4.8) acquire an extra dependence on $(\varkappa, \tilde{\varkappa})$ and, eventually, are expressed via sums $\sum_T \Pi_T(\alpha, \beta, \varkappa, \tilde{\varkappa})$ over pseudo-, 1- and 2-trees of $\hat{\Gamma}^\tau$ with

$$\Pi_T(\alpha, \beta, \varkappa, \tilde{\varkappa}) = \prod_{l \in \mathscr{L}_\alpha \setminus T} \alpha_l \prod_{l \in \mathscr{L}_\beta \cap T} \beta_l \prod_{\gamma \in F_\alpha} \varkappa_\gamma^{2L(\gamma \setminus T)} \prod_{\gamma \in F_\beta} \tilde{\varkappa}^{2L(\gamma \cap T)}. \tag{6.5}$$

Let us now turn to variables (5.3). To analyze the convergence of G_Δ it is efficient to rearrange the terms in (6.3) in accordance with some equivalence relation. We introduce an operation $F \to \bar{F}$ such that $F \subset \bar{F}$. The forest $\bar{F}$ is constructed as follows. Let $\gamma \in F_\alpha$ and let γ be the union of elements of F_α which strictly belong to γ. Note that $\mathscr{L}(\gamma)$ may be empty. Consider the nest $\{\gamma' \mid \gamma \neq \gamma' = \gamma \cap \gamma_l \cup \gamma,\ l = 1, \ldots, L\}$.
It includes $L(\gamma) - L(\gamma)$ different elements: $\gamma^1 \subset \gamma^2 \subset \cdots \subset \gamma^i \subset \cdots \subset \gamma^{L(\gamma)-L(\gamma)} \equiv \gamma$. For any $i = 1, 2, \ldots, L(\gamma) - L(\gamma) - 1$ add to F_α the UVI component of γ^i with the line $\mathscr{L}(\gamma^i \setminus \gamma^{i-1})$. After using in that way all the elements of the forest F_α we continue the construction procedure with the help of the nest

$$\{\gamma' \mid \gamma' = F_\alpha \cup \gamma_l, \qquad l = 1, \ldots, \bar{l}; \qquad \gamma' \neq F_\alpha\}.$$

IR part F_β of the forest is extended to $\mathscr{F}_\beta$ analogously with $\alpha \to \beta$, UV $\to$ IR; furthermore, one should replace the nest $\mathscr{N}$ by the "reversal" nest $\bar{\mathscr{N}} = \{\tilde{\gamma}_l \mid l = L \quad 1, \ldots, 1, 0\}$. At the final stage only subgraphs $\tilde{\gamma}_l \in \bar{\mathscr{N}}$ with $l \geqq \bar{l}$ are used.

Definition 6.1. Two forests F_1 and F_2 are equivalent if $\bar{F}_1 = \bar{F}_2$.

Now we turn to variables $(\boldsymbol{t}', \boldsymbol{\tau}')$ associated with a maximal forest $\mathscr{F}$ and introduced by (5.23). One easily obtains the relations between $(\boldsymbol{t}', \boldsymbol{\tau}')$ and $(\boldsymbol{t}, \boldsymbol{\tau})$

$$t_\gamma' = \begin{cases} t_{\sigma(\gamma)} \cdots t_{\sigma(\bar{\gamma})-1}, & \sigma(\gamma) < \sigma(\bar{\gamma}) \leqq \bar{l}; \\ (t_{\sigma(\bar{\gamma})} \cdots t_{\sigma(\gamma)-1})^{-1}, & \sigma(\bar{\gamma}) < \sigma(\gamma) \leqq \bar{l}; \\ t_{\sigma(\gamma)} \cdots t_{\bar{l}} \tau_{\bar{l}+1} \cdots \tau_{\sigma(\bar{\gamma})}, & \sigma(\gamma) \leqq \bar{l} < \sigma(\bar{\gamma}); \\ (t_{\sigma(\bar{\gamma})} \cdots t_{\bar{l}} \tau_{\bar{l}+1} \cdots \tau_{\sigma(\gamma)})^{-1}, & \sigma(\bar{\gamma}) \leqq \bar{l} < \sigma(\gamma); \\ \tau_{\sigma(\gamma)+1} \cdots \tau_{\sigma(\bar{\gamma})}, & \bar{l} < \sigma(\gamma) < \sigma(\bar{\gamma}); \\ (\tau_{\sigma(\bar{\gamma})+1} \cdots \tau_{\sigma(\gamma)})^{-1}, & \bar{l} < \sigma(\bar{\gamma}) < \sigma(\gamma) \end{cases}$$

$$\tau_\gamma' = \begin{cases} t_{\sigma(\bar{\gamma})} \cdots t_{\sigma(\gamma)-1}, & \sigma(\bar{\gamma}) < \sigma(\gamma) \leqq l; \\ (t_{\sigma(\gamma)} \cdots t_{\sigma(\bar{\gamma})-1})^{-1}, & \sigma(\gamma) < \sigma(\bar{\gamma}) \leqq \bar{l}; \\ t_{\sigma(\bar{\gamma})} \cdots t_{\bar{l}} \tau_{\bar{l}+1} \cdots \tau_{\sigma(\gamma)}, & \sigma(\bar{\gamma}) \leqq \bar{l} < \sigma(\bar{\gamma}); \\ (t_{\sigma(\gamma)} \cdots t_{\bar{l}} \tau_{\bar{l}+1} \cdots \tau_{\sigma(\bar{\gamma})})^{-1}, & \sigma(\gamma) \leqq \bar{l} < \sigma(\bar{\gamma}); \\ \tau_{\sigma(\bar{\gamma})+1} \cdots \tau_{\sigma(\gamma)}, & \bar{l} < \sigma(\bar{\gamma}) < \sigma(\gamma); \\ (\tau_{\sigma(\gamma)+1} \cdots \tau_{\sigma(\bar{\gamma})})^{-1}, & \bar{l} < \sigma(\gamma) < \sigma(\bar{\gamma}) \end{cases} \tag{6.6}$$

if γ is not maximal, and

$$t_\gamma' = \begin{cases} t_{\sigma(\gamma)} \dots t_{\bar{l}},\ \sigma(\gamma) \leqq \bar{l} \\ (\tau_{\bar{l}+1} \dots \tau_{\sigma(\gamma)})^{-1},\ \sigma(\gamma) > \bar{l}, \end{cases} \qquad \tau_\gamma' = \begin{cases} (t_{\sigma(\gamma)} \dots t_{\bar{l}})^{-1},\ \sigma(\gamma) \leqq \bar{l} \\ \tau_{\bar{l}+1} \dots \tau_{\sigma(\gamma)},\ \sigma(\gamma) > \bar{l} \end{cases} \tag{6.6'}$$

if γ is a maximal element of $\mathcal{F}$.

Now we decompose $\mathcal{F}_{\alpha(\beta)}$ as $\mathcal{F}^+_{\alpha(\beta)} \cup \mathcal{F}^-_{\alpha(\beta)}$. By definition, $\gamma \in \mathcal{F}_\alpha{}^+$ if either $\sigma(\gamma) < \sigma(\bar{\gamma})$ or $\sigma(\gamma) \leqq l$ and γ is maximal. Furthermore, $\gamma \in \mathcal{F}_\beta{}^+$ if either $\sigma(\bar{\gamma}) < \sigma(\gamma)$ or $\sigma(\gamma) > \bar{l}$ and γ is maximal. Thus, $\mathcal{F} = \mathcal{F}^+ \cup \mathcal{F}^-$ with $\mathcal{F}^\pm = \mathcal{F}_\alpha{}^\pm \cup \mathcal{F}_\beta{}^\pm$. One easily verifies that $\mathcal{F} = \bar{F}$ is always extended with elements of $\mathcal{F}^+$. Therefore, a term from an equivalence class of forests is characterized by $\mathcal{F}^-$ and a subset of $\mathcal{F}^+$. Hence, the contribution of the equivalence class to $\mathcal{R}^*G_\Delta$ is combined as follows

$$\mathcal{R}_{\mathcal{F}}{}^*G_\Delta = \prod_{\gamma\in\mathcal{F}_\alpha{}^+} (1 - M_\gamma) \prod_{\gamma\in\mathcal{F}_\alpha{}^-} (-M_\gamma) \prod_{\gamma\in\mathcal{F}_\beta{}^+} (1 - \tilde{M}_\gamma) \prod_{\gamma\in\mathcal{F}_\beta{}^-} (-\tilde{M})\, G_\Delta. \tag{6.7}$$

Let us now prove that the parametric integral $\mathcal{R}_{\mathcal{F}}{}^*G_\Delta$ regarded as a tempered distribution is absolutely convergent at $\lambda \in \mathcal{D}(\boldsymbol{n}, \tilde{\boldsymbol{n}})$ The crucial point is an appropriate choice of mixed representation (4.8) which corresponds to given set $\mathcal{L}_\alpha$ as was described in the previous Section. Now we take $\mathcal{L}_\alpha = \mathcal{L}(\mathcal{F}_\alpha)$. As a result the exponent H^r in the representation turns to be infinitely differentiable function of $(\boldsymbol{t}', \boldsymbol{\tau}')$. Thus, (6.7) takes the form

$$\begin{aligned} &\int_0^1 dt\, d\tau \prod_l t_l^{\nu_l+l-1} \prod_{l'} \tau_{l'}^{-\tilde{\nu}_{l'}+L-l'} \prod_{\gamma\in\mathcal{F}_\alpha{}^+} (I_{\varkappa_\gamma} - \mathcal{M}^{a\gamma}_{\varkappa_\gamma}) \\ &\times \prod_{\gamma\in\mathcal{F}_\alpha{}^-} (-\mathcal{M}^{a\gamma}_{\varkappa_\gamma}) \prod_{\gamma\in\mathcal{F}_\beta{}^+} (I_{\tilde{\varkappa}_\gamma} - \mathcal{M}^{\tilde{a}\gamma}_{\tilde{\varkappa}_\gamma}) \prod_{\gamma\in\mathcal{F}_\beta{}^-} (-\mathcal{M}^{\tilde{a}\gamma}_{\tilde{\varkappa}_\gamma}) \\ &\times \prod_{\gamma\in\mathcal{F}_\alpha} \varkappa_\gamma{}^{4\Re(\gamma)+a(\gamma)+R} \prod_{\gamma\in\mathcal{F}_\beta} \tilde{\varkappa}_\gamma{}^{4\hat{\Re}(\gamma)-a(\gamma)+R} f^r(\boldsymbol{x}, \boldsymbol{p}, \alpha, \beta, \varkappa, \tilde{\varkappa}) \end{aligned} \tag{6.8}$$

with $I_\varkappa g(\varkappa) = g(1)$. Now we use the relation

$$f^r(\dots \alpha, \beta, \dots) = \prod_{l=1}^{\bar{l}} \alpha_l{}^{-2} \prod_{l\in\mathcal{L}_\alpha} (\alpha_l)^2 f^r(\dots \alpha', \beta', \dots) \tag{6.9}$$

and representation (5.24) for $f^r(\dots \alpha', \beta', \dots)$

$$\begin{aligned} f^r(\boldsymbol{x}, \boldsymbol{p}, \boldsymbol{t}', \boldsymbol{\tau}', \varkappa, \tilde{\varkappa}) = &\prod_{\gamma\in\mathcal{F}_\alpha} (\varkappa_\gamma{}^2 t_\gamma')^{-[\omega(\gamma)/2]-L(\gamma)} \\ &\times \prod_{\gamma\in\mathcal{F}_\beta} (\tilde{\varkappa}^2_{\gamma'} \tau'_{\gamma'})^{-\Omega^\Gamma(\gamma')-L(\gamma')} Q(\boldsymbol{x}, \boldsymbol{p}, \boldsymbol{t}', \boldsymbol{\tau}, \varkappa\, \tilde{\varkappa}). \end{aligned} \tag{6.10}$$

Here the presence of the parameters $(\varkappa, \tilde{\varkappa})$ was taken into account. In (6.10) $Q \in C^\infty$ is a function depending on the combinations $\varkappa_\gamma{}^2 t_\gamma'$ and $\tilde{\varkappa}_\gamma{}^2 \tau_\gamma'$. Relations (6.6) and (6.6)' show that t_γ' or τ_γ' contain negative powers of $(\boldsymbol{t}, \boldsymbol{\tau})$ if, and only if $\gamma \in \mathcal{F}^-$. However, after operating by $\mathcal{M}_\varkappa$ or $\mathcal{M}_{\tilde{\varkappa}}$ the corresponding parameters $\varkappa_\gamma$ or $\tilde{\varkappa}_\gamma$ are set to zero. Therefore, representation (6.10) provides a factorization in $(\boldsymbol{t}, \boldsymbol{\tau})$. In other words, rhs of (6.10) after the action of M_γ or $\tilde{M}_\gamma$ $(\gamma \in \mathcal{F}^-)$ transforms into a product of powers of variables t_l and $\tau_{l'}$ multiplied by a C^∞-function.

Let us now establish UV convergence of the contribution $\mathcal{R}_{\mathcal{F}}{}^*G_\Delta$. It is necessary to show that $N_l \equiv \operatorname{Re} \deg t_l > -1$ ($\deg t_l$ is the power of t_l). With the help of (5.23) we have

$$\begin{aligned} N_l > \operatorname{Re} \nu_l - l - 1 + &\sum_{\gamma\in\mathcal{F}_\alpha:\sigma(\gamma)\leqq l<\sigma(\bar{\gamma})} N_\gamma{}^{\alpha,+} - \sum_{\gamma\in\mathcal{F}_\alpha:\sigma(\bar{\gamma})\leqq l<\sigma(\gamma)} N_\gamma{}^{\alpha,-} \\ + &\sum_{\gamma\in\mathcal{F}_\beta:\sigma(\bar{\gamma})\leqq l<\sigma(\gamma)} N_\gamma{}^{\beta,+} - \sum_{\gamma\in\mathcal{F}_\beta:\sigma(\gamma)\leqq l<\sigma(\bar{\gamma})} N_\gamma{}^{\beta,-}. \end{aligned} \tag{6.11}$$

Here $N_\gamma^{\alpha(\beta)\pm}$ are minimal for "+", and maximal for "−", powers of t_γ' or τ_γ' at $\gamma \in \mathcal{F}^\pm_{\alpha(\beta)}$ which occur in the integrand. It is understood in (6.11) that the inequalities $\sigma(\gamma) \leqq l < \sigma(\bar{\gamma})$, $\sigma(\bar{\gamma}) \leqq l < \sigma(\gamma)$ are replaced by $\sigma(\gamma) \leqq \bar{l}$ and $\sigma(\gamma) > \bar{l}$, respectively, if γ is a maximal element of $\mathcal{F}$. In the case of $\gamma \in \mathcal{F}^-$ (6.1), (6.8)—(6.10) and the property of the function Q lead to

$$-N_\gamma^{\alpha(\beta),-} \geqq \operatorname{Re} \nu(\gamma) - L(\gamma). \tag{6.12}$$

If $\gamma \in \mathcal{F}^+$ then the formula for a remainder

$$(I_\varkappa - \mathcal{M}_\varkappa^k)\, g(\varkappa) = \frac{1}{k!} \int\limits_0^1 d\varkappa (1 - \varkappa)^k\, g^{(k+1)}(\varkappa) \tag{6.13}$$

should be used. Thus, the operating of $I_\varkappa - M_\varkappa^k$ is, in fact, equivalent to the action of $(\partial/\partial\varkappa)^{k+1}$ as far as the power counting is carried out. Therefore, we have

$$N_\gamma^{\alpha(\beta),+} \geqq -\operatorname{Re} \nu(\gamma) + L(\gamma). \tag{6.14}$$

Now we rearrange the terms in (6.11) with the help of

$$\sum_{\sigma(\gamma)\leqq l<\sigma(\bar{\gamma})} r_\gamma = \sum_{\sigma(\gamma)\leqq l} r_\gamma - \sum_{\sigma(\gamma),\sigma(\bar{\gamma})\leqq l} r_\gamma,\ \gamma \in \mathcal{F}_\alpha;$$

$$\sum_{\sigma(\bar{\gamma})\leqq l<\sigma(\gamma)} r_\gamma = \sum_{\gamma:\sigma(\bar{\gamma})\leqq l} r_\gamma - \sum_{\sigma(\gamma),\sigma(\bar{\gamma})\leqq l} r_\gamma,\ \gamma \in \mathcal{F}_\beta;$$

$$\sum_{\gamma:\sigma(\bar{\gamma})\leqq l} r_\gamma = \sum_{\gamma:\sigma(\gamma)\leqq l}\ \sum_{\gamma':\bar{\gamma}'=\gamma} r_{\gamma'}. \tag{6.15}$$

Finally, we come to $N_l > -1$ and this implies UV convergence. As to IR convergence, it is demonstrated analogously by means of relations (6.1), (6.8)—(6.10), (6.13), (6.15).

6.3. Analytic extension

Let us now prove that absolutely convergent integrals (6.2) at different $(\boldsymbol{n}, \tilde{\boldsymbol{n}})$ may be obtained from each other by analytic extension procedure. Note that it is always possible to connect two points $\boldsymbol{\lambda} \in \mathcal{D}(\boldsymbol{n}, \tilde{\boldsymbol{n}}) \equiv \mathcal{D}$ and $\boldsymbol{\lambda}' \in \mathcal{D}(\boldsymbol{n}', \tilde{\boldsymbol{n}}') \equiv \mathcal{D}'$ by a path with the following property: when going by this path no more than one parameter n_γ or $\tilde{n}_\gamma$ may change simultaneously. Specifically, consider only two neighbouring domains with $n'_{\gamma^0} = n_{\gamma^0} + 1$; $n_\gamma' = n_\gamma$, $\gamma \neq \gamma^0$, $\gamma \in \mathcal{U}$ and $\tilde{n}_\gamma' = \tilde{n}_\gamma$, $\gamma \in \mathcal{J}$. There exist $\varepsilon > 0$ and $\delta > 0$ such that along the path connecting these domains the following inequalities hold $|\operatorname{Re} \nu(\gamma^0) + n_{\gamma^0} + 1 - [\omega(\gamma^0)/2]| \leqq \varepsilon$ and

$$\left|\operatorname{Re} \nu(\gamma) + n_\gamma + \frac{1 \mp 1}{2} - [\omega(\gamma)/2]\right| \geqq \delta, \qquad \gamma \neq \gamma^0$$

as well as analogous IR inequalities. Note that we may consider a sufficiently small path and provide the relation $\varepsilon \ll \delta$. Then we choose the domain $\mathcal{D}_\varepsilon \supset \mathcal{D} \cup \mathcal{D}'$ with $\mathcal{D} = \mathcal{D}(\boldsymbol{n}, \tilde{\boldsymbol{n}}) \cap \mathcal{D}_\varepsilon$, $\mathcal{D}' = \mathcal{D}(\boldsymbol{n}', \tilde{\boldsymbol{n}}') \cap \mathcal{D}_\varepsilon$ which includes this path connecting $\mathcal{D}$ and $\mathcal{D}'$ through the boundary $\operatorname{Re} \nu(\gamma^0) + n_{\gamma^0} + 1 - [\omega(\gamma^0)/2] = 0$.

Let us show that the analytic extension leads to the change $n_0 = n_{\gamma^0} \to n_0 + 1$. The method used here generalizes as much as possible the well-known analytic extension of the functional x_+^λ [18].

1) We first split integral (6.2) at $\boldsymbol{\lambda} \in \mathcal{D} \subset \mathcal{D}(\boldsymbol{n}, \tilde{\boldsymbol{n}})$ into the sum of two integrals $G_1 + G_2$ in accordance with decomposition of the whole integration domain into

$$\Delta_1 = \{\alpha \mid \alpha_l \leqq 1, l \in \gamma^0\} \qquad \text{and} \qquad \Delta_2 = \{\alpha \mid \exists\, l \in \gamma^0 : \alpha_l > 1\}$$

2) Let us show that the contribution of any sector $\Delta_{\pi,\bar{l}}$ to G_2 naturally extends to $\mathcal{D}'$. The inequality $N_l > -1$ follows from relations (6.11), (6.12), the latter being a consequence of (6.1). On the path under consideration only one of these inequalities changes according to $n_0 \to n_0 + 1$. However, this is compensated by contributions from other relations (6.1) which hold up to order δ (see above). The fact is that $N_\gamma^{\alpha,\pm}$ can not be the only contribution to (6.11) from four sums in rhs. Indeed, suppose that it is not true. Then $\sigma(\gamma) \leqq l < \sigma(\bar{\gamma}^0)$ and for $\gamma \neq \gamma^0$, $\gamma \in \mathcal{F}_\alpha$ either $\sigma(\gamma) > l$ or $\sigma(\bar{\gamma}) \leqq l$. If $\bar{\gamma} = \gamma^0$ then the inequality $\sigma(\gamma) > l$ can not take place because $\sigma(\bar{\gamma}) = \sigma(\gamma^0) \leqq l$. Hence $\sigma(\gamma) \leqq l$. Analogously, one obtains $\sigma(\gamma) \leqq l \;\forall\, \gamma \in \mathcal{F}, \gamma \subset \gamma^0$, i.e. $\gamma^0 \subset \gamma_l \subset \gamma_{L_0}$. However $l \leqq \bar{l}$, but the sectors $\Delta_{\pi,\bar{l}}$ at $\gamma^0 \subset \gamma_{\pi(L_0)}$ do not contribute to G_2. Thus, the assumption does not hold. As to $N_\gamma^{\alpha,-}$, the corresponding estimates are valid both in $\mathcal{D}$ and $\mathcal{D}'$.

3) Let us represent $-M^{a^{\gamma^0}}$ as $-M^{a_0+2} + M^{(a_0+1)} + M^{(a_0+2)}$. Here $a_0 = a^{\gamma^0}$, and $M^{(k)}$ picks out only k-th term of Taylor expansion. Note that only one of the last two operators gives a non-zero contribution (P_l are homogeneous). Suppose that this is $M^{(a_0+2)}$. The decomposition mentioned above leads to the representation $G_1 = G_1{}^1 + G_1{}^2$ where $G_1{}^1$ corresponds to $\mathcal{R}^*$ written for domain $\mathcal{D}(\boldsymbol{n}', \tilde{\boldsymbol{n}}')$ (i.e. $a_0 \to a_0 + 2$ in comparison with $\mathcal{D}(\boldsymbol{n}, \tilde{\boldsymbol{n}})$). As to $G_1{}^2$, it is associated with the following contribution to $\mathcal{R}^*$

$$\mathcal{R}^*_{\gamma^0} = M^{(a_0+2)}_{\gamma^0} \sum_{F \ni \gamma^0} \prod_{\gamma^0 \neq \gamma \in F_\alpha} (-M_\gamma{}^{a^\gamma}) \prod_{\gamma \in F_\beta} (-M_\gamma{}^{a^\gamma}).$$

4) The integral $G_1{}^1$ is absolutely convergent in $\mathcal{D}$ and extends naturally to $\mathcal{D}'$: this proposition may be proved as the analogous proposition at the second step.

5) Let us extend $G_1{}^2$ to $\mathcal{D}'$. For convenience the numbering of lines begins with $\mathcal{L}(\gamma^0)$. Decompose the integration region over $\boldsymbol{\alpha}^0 = \{\alpha_l \mid l \in \gamma^0\}$ into (usual) H-sectors

$$\Delta^{\pi_0} = \{\boldsymbol{\alpha}^0 \mid \alpha_{\pi_0(1)} \leqq \cdots \leqq \alpha_{\pi_0(L_0)}\},$$

each of them being characterized by a permutation π_0 of numbers $1, \ldots, L_0 \equiv L(\gamma^0)$. Let us represent Δ^{π_0} as $\Delta_1{}^{\pi_0} \cup \Delta_2{}^{\pi_0}$ where $\Delta^{\pi_0}_{1,2}$ are defined by the inequalities $\alpha_{\pi(L_0)} \leqq 1$ and $\alpha_{\pi(L_0)} > 1$ respectively. Note that $\Delta_1{}^{\pi_0}$ gives contribution to G_1, $G_1{}^1$, $G_1{}^2$ and $\Delta_2{}^{\pi_0}$ contributes to G_2. Without loss of generality consider only the sector associated with identical permutation. Introduce H-variables $\boldsymbol{t}^0 = (t_1{}^0, \ldots, t^0_{L_0})$ as follows $\alpha_l = t_l{}^0 \ldots t^0_{L_0}$, $l = 1, \ldots, L_0$. The region of parameters $\boldsymbol{\alpha}^1 = \{\alpha_l \mid l \notin \gamma^0\}$ is decomposed into sectors $\Delta_{\pi^1,\bar{l}}$ associated with permutations π^1 of numbers $L_0 + 1, \ldots, L$ and with numbers $\bar{l}$ which separate α and β-variables. The sector variables $(\boldsymbol{t}^1, \boldsymbol{\tau}^1)$ are expressed in terms of $(\boldsymbol{\alpha}^1, \boldsymbol{\beta}^1)$ by the same relations as in the case of $(\boldsymbol{t}, \boldsymbol{\tau})$ and $(\boldsymbol{\alpha}, \boldsymbol{\beta})$ (see Sect. 5). One has $\boldsymbol{\tau}^1 = \boldsymbol{\tau}$. The variables $\boldsymbol{t}$ are related with $\boldsymbol{t}^0$ and $\boldsymbol{t}^1$ as follows:

$$t_l = t_l{}^0,\ l = 1, \ldots, L_0 - 1; \qquad t_l = t^0_{L_0}/t^1_{L_0+1} \ldots t^1_{L_0+\bar{l}},\ l = L_0;$$

$$t_l = t_l{}^1,\ l = L_0 + 1, \ldots, L_0 + \bar{l}.$$

To analyze the convergence of $G_1{}^2$ one should (as in Subsect. 6.2) decompose the contribution to $\mathcal{R}^*$ in accordance with the equivalence relation. Since $\varkappa_{\gamma^0} = 0$, after the action of M_{γ^0}, the factorization in $(\boldsymbol{t}, \boldsymbol{\tau})$ provides a factorization in $(\boldsymbol{t}^0, t^1, \boldsymbol{\tau}^1)$, the former being obtained with the help of the factorization in $(\boldsymbol{t}', \boldsymbol{\tau}')$ (see Subsect. 6.2).

Performing the double power counting one has the conditions

$$N_l{}^0 = \operatorname{Re}\deg t_l{}^0 > -1,\ l = 1, \ldots, L_0 - 1;$$

$$N_l{}^1 = \operatorname{Re}\deg t_l{}^1 > -1,\ l = L_0 + 1, \ldots, L_0 + \bar{l},\ \tilde{N}_{l'} = \operatorname{Re}\deg \tau_{l'} > -1,$$

$$l' = L_0 + \bar{l} + 1, \ldots, L$$

which are valid both in $\mathcal{D}$ and $\mathcal{D}'$. As to the variable $t^0_{L_0}$, it enters into the integrand in some power. A straightforward calculation shows that the integral

$$\int_0^1 dt^0_{L_0} (t^0_{L_0})^{\nu_{L_0} - [\omega(\gamma^0)/2] + n_0}$$

is exhibited. Its analytic extension from $\mathcal{D}$ to $\mathcal{D}'$ looks like

$$-\int_1^\infty dt^0_{L_0} (t^0_{L_0})^{\nu_{L_0} - [\omega(\gamma^0)/2] + n_0}.$$

Therefore, the whole integral $G_1{}^2$ extends to D'. Here it takes the form of an integral over Δ_2 which is the union of the sectors $\Delta_2{}^{\pi_0}$

6) Now we combine the obtained expression with G_2 and get an integral over Δ_2. Note that inside the integrand there is the operator $\mathcal{R}^*$ corresponding to $\mathcal{D}(\boldsymbol{n}', \tilde{\boldsymbol{n}}')$.

7) Finally, we add the integral G_1 analytically continued to $\mathcal{D}'$ and obtain representation (6.2), $\mathcal{R}^*$ being now associated with $\mathcal{D}(\boldsymbol{n}', \tilde{\boldsymbol{n}}')$. Note that this representation turns to be absolutely convergent in all this region, and not only in $\mathcal{D}'$.

6.4. Dimensionally regularized Feynman amplitudes

It is easy to generalize the results of Subsect. 5.3 and 6.1–6.3 to the case when both the analytic and the dimensional [32] regularizations are involved. If the latter is introduced then the additional factor $(D(\alpha))^\varepsilon$ appears in (1.3). Here $\varepsilon = (4 - d)/2$ and d is the space-time dimension. With the help of the factorization of $D \equiv A^1$ (see Appendix B) one has that the assertion of theorem 5.1 remains the same, but, formulae (5.14)–(5.16) should be changed. In particular, inequalities (5.14) transform to

$$\begin{aligned} &\operatorname{Re}\nu(\gamma) + \mathfrak{N}(\gamma)\operatorname{Re}\varepsilon > \bar{\omega}(\gamma), \qquad \gamma \in \mathcal{U}; \\ &\operatorname{Re}\nu(\gamma) + \mathfrak{N}(\Gamma/\bar{\gamma})\operatorname{Re}\varepsilon < -\Omega^\Gamma(\gamma), \qquad \gamma \in \mathcal{J}. \end{aligned} \tag{6.16}$$

Similarly, (5.15) and (5.16) are modified. As in Sect. 5, one proves that the initial analyticity domain of analytically and dimensionally regularized FA $G_\Gamma(\boldsymbol{p}; \boldsymbol{\lambda}, \varepsilon)$ is not empty: instead of parameters (5.17), (5.22) it is necessary to use

$$\lambda_l{}^\Gamma = \frac{d}{2}\left(1 + \delta - \Big(\sum_{T \in T_l{}^{[1]}} 1\Big) \Big/ \Big(\sum_{T \in T^{[1]}} 1\Big)\right) - 1 + a_l/2. \tag{6.17}$$

Theorem 6.1 remains true as well. One should insert the factor D^ε into (6.2). The degrees of subtracting operators $M_\gamma^{a^\gamma}$ and $\tilde{M}_\gamma^{\tilde{a}^\gamma}$ in (6.3) are now defined as follows

$$a_\gamma = \omega(\gamma) + 2[-\mathrm{Re}\,\nu(\gamma) - \mathfrak{N}(\gamma)\,\mathrm{Re}\,\varepsilon],$$

$$\tilde{a}_\gamma = \tilde{\omega}(\gamma) + 1 + 2[\mathrm{Re}\,\nu(\gamma) + \mathfrak{N}(\Gamma/\tilde{\gamma})\,\mathrm{Re}\,\varepsilon].$$

Let us now use the analytic regularization as auxiliary one. Suppose that there is no domain in complex ε-plane in which the α-representation of dimensionally regularized FA (DRFA) $G_\Gamma(\boldsymbol{p}; \varepsilon)$ is absolutely convergent. Introduce then the analytic regularization. In the case of a graph without massless tadpoles domain (6.16) is not empty. Therefore, the analytically and dimensionally regularized FA $G_\Gamma(\boldsymbol{p}; \lambda, \varepsilon)$ is unambiguosly defined in this domain from which it extends to a meromorphic function. In particular, one may continue FA to the point $(0, \varepsilon)$, i.e. remove the analytic regularization, if the following inequalities hold $\mathfrak{N}(\gamma)\,\varepsilon \neq \overline{\omega}(\gamma) - n$, $\gamma \in \mathcal{U}$, $\mathfrak{N}(\Gamma/\tilde{\gamma})\,\varepsilon \neq -\Omega^\Gamma(\gamma) + n$, $\gamma \in \mathcal{J}$; $n = 0, 1, \ldots$

Thus, we have a natural definition of DRFA possessing both UV and IR divergences.

Definition 6.2. If there are no massless tadpoles in Γ then the corresponding DRFA considered in the region

$$\big\{\varepsilon \in \mathbf{C} \mid ([\omega(\gamma)/2] - n_\gamma - 1)/\mathfrak{N}(\gamma) < \mathrm{Re}\,\varepsilon < \big([\omega(\gamma)/2] - n_\gamma\big)/\mathfrak{N}(\gamma),$$

$$\gamma \in U,\ \mathfrak{N}(\gamma) \neq 0;\ \big(\tilde{n}_\gamma - \Omega^\Gamma(\gamma)\big)/\mathfrak{N}(\Gamma/\tilde{\gamma}) < \mathrm{Re}\,\varepsilon < \big(\tilde{n}_\gamma - \Omega^\Gamma(\gamma) + 1\big)/\mathfrak{N}(\Gamma/\tilde{\gamma}),$$

$$\gamma \in \mathcal{J},\ \mathfrak{N}(\Gamma/\tilde{\gamma}) \neq 0\big\}$$

is defined by the absolutely convergent integral

$$\int d\alpha \mathcal{R}^* D^\varepsilon f_\Gamma \tag{6.18}$$

where $\mathcal{R}^*$ is given by (6.3) with

$$a^\gamma = \omega(\gamma) + 2[-\mathfrak{N}(\gamma)\,\mathrm{Re}\,\varepsilon], \qquad \tilde{a}^\gamma = \tilde{\omega}(\gamma) + 1 + 2[\mathfrak{N}(\Gamma/\tilde{\gamma})\,\mathrm{Re}\,\varepsilon].$$

It should be noted that the absolutely convergent α-representation allows to accomplish explicitly the analytic continuation procedure in the definition of arbitrary DRFA given in ref. [30] (an independent definition was formulated in ref. [31] by means of the method based on the Mellin representation). This definition makes it possible to justify the formal manipulations which are of current use in practical calculations. For example, one may apply integration by parts in dimensionally regularized Feynman integrals [33], differentiation in external momenta, decomposition of an integral into a sum of integrals etc.

In fact, a definition of DRFA which is equivalent to that of ref. [30] may be formulated as follows. DRFA corresponding to a graph with a massless tadpole is zero, otherwise, it is defined by absolutely convergent integral (6.18).

Finally, note that, probably, it is more natural to regard DRFA as a function in p-space rather than a tempered distribution. From this point of view DRFA do not have poles associated with superficial IR divergences. Indeed, when considering conditions on the analyticity domain and on the position of IR poles, one may take into account only subgraphs γ with essential complements $\tilde{\gamma}$.

7. Conclusion: R^*-operation

As was shown in Sect. 6 the explicit analytic extension of analytically and/or dimensionally regularized FA is accomplished by the operator $\mathcal{R}^*$ (See (1.6), (6.3)) which is a generalization of the operator $\mathcal{R}$ (see (1.5)) possessing the structure of BPHZ R-operation. The R^*-operation [23] turns to be an analogous generalization of the dimensional renormalization [32]. For example, R^* may be represented as

$$R^* = \sum_{F=F_\alpha \cup F_\beta} \prod_{\gamma \in F_\alpha} (-K_\gamma) \prod_{\gamma \in F_\beta} (-\tilde{K}_\gamma) \tag{8.1}$$

where the sum goes over (generalized) forests of the graph and K_γ is the operator which picks out the UV-pole part of Laurent expansion in $\varepsilon = (4 - d)/2$ of FA G_γ. Similarly, $\tilde{K}_\gamma$ picks out the IR-pole part of the expansion.

For practical applications the R^*-operation written in the counterterm form [23] is needed

$$R^* = \sum_{F=F_\alpha \cup F_\beta} \Delta(F_\alpha)\, \tilde{\Delta}(F_\beta) = {}'R^* + \Delta(\Gamma) + \tilde{\Delta}(\Gamma). \tag{8.2}$$

Here the sum runs over forests $F = F_\alpha \cup F_\beta$, and $F_\alpha = \bigcup_i \gamma_i$, $F_\beta = \bigcup_j \gamma_j'$ consist of UVI (IRI) VU(IR)-divergent (i.e. $\omega(\gamma_i)$, $\tilde{\omega}(\gamma_j') \geqq 0$) subgraphs γ_i and γ_j' which have not lines in common. The counterterm operators are defined as follows:

$$\Delta(\gamma) = \tilde{\Delta}(\gamma') = 1 \quad \text{if} \quad \mathcal{L}(\gamma) = \mathcal{L}(\gamma') = \emptyset;$$

$$\Delta(F_\alpha) = \prod_i \Delta(\gamma_i), \qquad \tilde{\Delta}(F_\beta) - \prod_j \tilde{\Delta}(\gamma_j').$$

The operating of $\Delta(\gamma)$ reduces to calculating of UV counterterm and inserting it into the diagram Γ/γ. Similarly, the operating of $\tilde{\Delta}(\gamma')$ reduces to calculating of IR counterterm and inserting it into the diagram $\Gamma \setminus \gamma'$. Note that the latter may be disconnected. It should be emphasized that IR counterterms are local in p-space and, therefore, essentially non-local in x-space. Both R^*-operation itself and methods of calculating UV and IR counterterms will be presented in detail in further publications.

Thus, R^*-operation removes both UV and IR poles. A generalization of Bogoliubov-Parasiuk theorem in the framework of the dimensional renormalization takes place: $R^* G_\Gamma(\boldsymbol{p}, \varepsilon)$ is finite at $\varepsilon \to 0$ for every Γ. The proof based on the method of this paper will be presented elsewhere.

Note that a number of works [24] employed successfully R^*-operation both in calculations and in studying various asymptotic expansion of Feynman integrals has been appeared. It should be stressed that the method of the present paper turns to be very effective for justifying the results which concern the R^*-operation because UV and IR divergences are simultaneously involved. For the same reason this method will be applied in the following paper in studying renormalization and operator product expansion in theories including massless particles. The author hopes that the method will find further applications.

The author is grateful to M. C. Polivanov for support and to S. A. Anikin and K. G. Chetyrkin for useful remarks.

Appendix A. Factorization in *H*-variables

By introducing the sector variables all the polylinear forms are factorized: a product of these variables in some powers is picked out with sufficient accuracy. We begin with H-sectors. First, let us prove the following relation

$$\Pi_T(t, \tau) = \prod_l t_l^{\mathfrak{N}(\hat{\gamma}_l^r) + \mathrm{def}_T(\hat{\gamma}_l^r) - \mathfrak{N}(\hat{\gamma}_l^r \cap T)} \prod_{l'} \tau_{l'}^{\mathfrak{M}^{\Gamma^r}(\gamma'_{l'}) + \mathrm{def}_T(\hat{\gamma}^r_{l'-1}) - \mathfrak{N}(\hat{\gamma}^r_{l'-1}) + 1 - n} \tag{A.1}$$

where T is a pseudo- or n-tree, and $\mathfrak{N}(\gamma \cap T) = 0$ in the latter case. It is understood that $l = 1, \ldots, \bar{l}$; $l' = \bar{l} + 1, \ldots, L$, and γ_l, $\gamma'_{l'}$ stand for the subgraphs composed of lines $\{1, \ldots, l\}$ and $\{l', \ldots, L\}$, respectively. Note that $\tilde{\gamma}'_{l'} = \gamma_{l'-1}$.

To obtain (A.1) let us represent Π_T as follows

$$\Pi_T(t, \tau) = \prod_l t_l^{L(\hat{\gamma}_l^r \cap T)} \prod_{l'} \tau_{l'}^{L(\hat{\gamma}'^r_{l'} \cap T) - r}$$

(we take into account that $L(\hat{\gamma}^r) = L(\gamma) + r$). Then we use the equalities

$$L(\hat{\gamma}'^r \cap T) - r = L(T) - L(\tilde{\hat{\gamma}}'^r) - r = L(T) - L(\tilde{\hat{\gamma}}'^r \cap T),$$

$$L(\hat{\gamma}^r \cap \tilde{T}) = L(\hat{\gamma}^r) - L(\hat{\gamma}^r \cap T).$$

If T is n-tree then $L(T) = L + r - \mathfrak{N}(\hat{\Gamma}^r) + 1 - n$, and for pseudo-tree it is necessary to set $n = 0$ in rhs of this formula. By applying the relations

$$\mathfrak{N}(\hat{\gamma}^r \cap T) = L(\hat{\gamma}^r \cap T) - V(\hat{\gamma}^r \cap T) + c(\hat{\gamma}^r \cap T),$$

$$\mathfrak{N}(\hat{\gamma}^r) = L(\hat{\gamma}^r) - V(\hat{\gamma}^r) + c(\hat{\gamma}^r), \qquad V(\hat{\gamma}^r \cap T) = V(\hat{\gamma}^r)$$

we find that

$$L(\hat{\gamma}^r \cap T) = L(\hat{\gamma}^r) + \mathfrak{N}(\hat{\gamma}^r \cap T) - \mathrm{def}_T\, \hat{\gamma}^r. \tag{A.2}$$

Inserting (A.2) into the powers of t_l and $\tau_{l'}$ yields (A.1).

All the polylinear forms are represented as the sums $\sum_T \Pi_T$ over pseudo-, 1-, and 2-trees of the graph $\hat{\Gamma}^r$. To establish the factorization one should choose maximal common powers of t_l and $\tau_{l'}$. In other words, it is necessary to construct the pseudo, 1- or 2-tree which provides the minimum of $\mathrm{def}_T\, \gamma - \mathfrak{N}(\gamma \cap T)$ with a sufficient accuracy. While performing the minimization procedure it will be supposed that all the lines $\hat{\mathscr{L}}^r$ belong to this pseudo-, 1- or 2-tree from the very beginning. As a result any form $F^r = A^r, A_i^{r,j}$, $\ldots, K^r_{l_1 l_2}$ is represented as follows

$$F^r(t, \tau) = \prod_l t_l^{\mathfrak{N}(\hat{\gamma}_l^r) + d_{F^r}(\gamma_l)} \prod_{l'} \tau_{l'}^{\mathfrak{M}^{\hat{\Gamma}^r}(\gamma'_{l'}) + d_{F^r}(\gamma_{l'-1}) + 1 - n} P_{F^r}(t, \tau) \tag{A.3}$$

where P_{F^r} is a polynomial.

Let us prove the factorization property of $A^r(\alpha, \beta)$ with the use of (A.1) and representation (4.10). To choose the tree $T_A \in \hat{T}^{[1]}$ which provides the minimum of the defects $\mathrm{def}_T\, (\hat{\gamma}_l^r)$ we apply the standard graph-theoretical procedure: the lines $1, \ldots, l, \ldots, L$ are consequently included into T_A if circuits are not generated. It is easy to see that $\mathrm{def}_T\, (\hat{\gamma}_l^r) = 0$ for all $l = 1, \ldots L$ and $T = T_A$. The non-negativity of the defects implies the minimum. Hence

$$A^r(\boldsymbol{t}, \boldsymbol{\tau}) = \prod_l t_l^{\mathfrak{N}(\hat{\gamma}_l^r)} \prod_{l'} \tau_{l'}^{\mathfrak{M}^{\hat{\Gamma}^r}(\gamma'_{l'})} \big(1 + P_A^r(\boldsymbol{t}, \boldsymbol{\tau})\big) \tag{A.4}$$

where P_A^r is a polynomial with non-negative coefficients. By using (4.10) let us now prove the following factorization

$$A_i^{r,j}(\boldsymbol{t}, \boldsymbol{\tau}) = \prod_l t_l^{\mathfrak{N}(\hat{\gamma}_l^r)+d_i^{r,j}(\gamma_l)} \prod_{l'} \tau_{l'}^{\mathfrak{M}^{\hat{\Gamma}^r}(\gamma'_{l'})+d_i^{r,j}(\gamma_{l'-1})} P_i^{r,j}(\boldsymbol{t}, \boldsymbol{\tau}). \tag{A.5}$$

Here $P_i^{r,j}$ is a polynomial that does not involve $t_1, \ldots, \tau_L$ as common factors, and $d_i^{r,j}(\gamma) = 1$ if the vertex v_i is connected with $v_j \in \mathcal{V}^r$ in $\hat{\gamma}^r$ but every such path goes through some other vertex $v_{j'} \in \mathcal{V}^r (j' \neq j)$, and $d_i^{r,j}(\gamma) = 0$ elsewhere.

The tree $T(\gamma)$ which provides the minimum of $\text{def}_T\, \gamma$ is constructed as follows.

a) Suppose that v_i is not connected with $\hat{v}^r$ in $\hat{\gamma}^r$, hence $d_i^{r,j}(\gamma) = 0$. To construct $T(\gamma)$ let us first use the lines of $\hat{\gamma}^r$: we include them into the tree (in an arbitrary order) if the inclusion does not generate a circuit. Then we construct a path in $\hat{\Gamma}^r/\gamma$ that does not go through $v_{j'} \in \mathcal{V}^r$ $(j' \neq j)$ and connects v_j with v_i (or with the vertex which replaces, after reduction, some other vertices, including v_i). Then we include the lines of this path into the tree. Finally, the rest lines of Γ are similarly used.

b) Suppose that the vertex v_i is connected in $\hat{\gamma}^r$ with v_j by a path which does not go through other vertices of $\mathcal{V}^r$, hence $d_i^{r,j}(\gamma) = 0$. First, we include this path into the tree. After that we use the other lines of γ for the construction of the tree. Then the lines of $\tilde{\gamma}$ are analogously used.

c) Let $d_i^{r,j}(\gamma) = 1$, i.e. the vertex v_i is connected with $v_{j'} \in \mathcal{V}^r$ in $\hat{\gamma}^r$ and, therefore, v_i is connected with v_j through $\hat{v}^r$. Thus, any path connecting v_i and v_j goes through some $v_{j''} \in \mathcal{V}^r$ $(j'' = j)$. Let us use all the lines of γ (in an arbitrary order) to construct a tree for every connectivity component of γ. As a result a path connecting v_i and some $v_{j'} \in \mathcal{V}^r$ $(j' \neq j)$ arises. Let us consider the component $\hat{\gamma}_0^r$ of $\hat{\gamma}^r$ in which v_i is connected with $v_{j'}$. Remove a line l_0 from the path involved. Include then into $T(\gamma)$ all the other lines belonging to the trees of the components of $\hat{\gamma}^r$. Furthermore, we construct a path in $\hat{\Gamma}^r/(\hat{\gamma}_0^r \setminus \{l_0\})$ which connects v_i and v_j and does not go through other vertices of $\mathcal{V}^r$. We include this path into $T(\gamma)$. Finally, all the other lines of T are used as above.

By considering separately every case it is easy to verify that the tree $T(\gamma)$ provides the minimum $d_i^{r,j}(\gamma)$ of the defect. This implies factorization (A.5).

The factorization of $A_{ii'}^r$ and $U_{jj'}^r$ may be established analogously by the minimization procedure. However, it is sufficient to apply (A.5). Furthermore, relation (3.5), as well as $U_{jj'}^r = A_{j'}^{r-1,j}$ ($\mathcal{V}^{r-1}$ is obtained from $\mathcal{V}^r$ by removing the vertex $v_{j'}$), should be used. By calculating the changes of $\mathfrak{N}(\hat{\gamma}_l^r)$ and $\mathfrak{M}^{\hat{\Gamma}^r}(\gamma'_{l'})$, in accordance with the change of $\mathcal{V}^r$, one comes to the following factorizations

$$A_{ii'}^r(\boldsymbol{t}, \boldsymbol{\tau}) = \prod_l t_l^{\mathfrak{N}(\hat{\gamma}_l^r)+d_{ii'}^r(\gamma_l)} \prod_{l'} \tau_{l'}^{\mathfrak{M}^{\hat{\Gamma}^r}(\gamma'_{l'})+d_{ii'}^r(\gamma_{l'-1})-1} P_{ii'}^r(\boldsymbol{t}, \boldsymbol{\tau}), \tag{A.6}$$

$$U_{jj'}^r(\boldsymbol{t}, \boldsymbol{\tau}) = \prod_l t_l^{\mathfrak{N}(\hat{\gamma}_l^r)+d^{r,jj'}(\gamma_l)-1} \prod_{l'} \tau_{l'}^{\mathfrak{M}^{\hat{\Gamma}^r}(\gamma'_{l'})+d^{r,jj'}(\gamma_{l'-1})} P^{r,jj'}(\boldsymbol{t}, \boldsymbol{\tau}). \tag{A.7}$$

Here $P_{ii'}^r$ and $P^{r,jj'}$ are polynomials without any $t_1, \ldots, \tau_L$ as a common factor. In (A.6) $d_{ii'}^r(\gamma) = 2$ if v_i and v_i are not connected in $\hat{\gamma}^r$ by a path which does not go through the vertices of $\mathcal{V}^r$ and both v_i and $v_{i'}$ are connected with $\hat{v}^r$; $d_{ii'}^r(\gamma) = 0$ if v_i and $v_{i'}$ are not connected in $\hat{\gamma}^r$ with $\hat{v}^r$, and $d_{ii'}^r(\gamma) = 1$ elsewhere.

Let us now establish the factorization of the forms associated with the linear (in ξ) part of the exponent.

The definition of $Q^r_{l_0}$ shows that it is a sum of products (A.1) with coefficients depending on x. The sum runs over the set $\hat{T}^{[0]}_{l_0}$ of pseudo-trees of $\hat{\Gamma}^r$ with a circuit going through l_0 and $\hat{v}^r$. As to $-\mathfrak{N}(\gamma^r \cap T)$, it approaches the minimum $-\psi^r_{l_0}(\gamma)$. Here $\psi^r_{l_0}(\gamma) = 1$ if there exists a circuit in γ^r going through l_0 and $\hat{v}^r$, and $\psi^r_{l_0}(\gamma) = 0$ elsewhere. To construct the pseudo-tree $T(\gamma)$ which provides the minimum it is sufficient to include any circuit of γ^r. In the case of $\mathrm{def}_T(\gamma_l^r)$ it suffices to use its non-negativity. Thus, we have the factorization

$$Q^r_{l_0}(\boldsymbol{x}, \boldsymbol{t}, \boldsymbol{\tau}) = \prod_l t_l^{\mathfrak{N}(\hat{\gamma}_l^r)-\psi^r_{l_0}(\gamma_l)} \prod_{l'} \tau_{l'}^{\mathfrak{M}^{\hat{\Gamma}^r}(\gamma'_{l'})+1-\psi^r_{l_0}(\gamma_{l'-1})} P^{r,Q}_{l_0}(\boldsymbol{t}, \boldsymbol{\tau}). \tag{A.8}$$

The polylinear form $Y^r_{l_0}$ is represented as the sum of (A.1) over the set of trees $\hat{T}^{[1]}_{l_0}$ in which some vertex $v_i \in \mathcal{V}^k \setminus \mathcal{V}^r$ is connected with $\hat{v}^r$ through the line l_0. It is sufficient to use the following inequality $\mathrm{def}_T(\gamma^r) \geqq y^r_{l_0}(\gamma)$. Here $y^r_{l_0}(\gamma) = 1$ if $\hat{v}$ and all the vertices $\mathcal{V}^k \setminus \mathcal{V}^r$ belong to the same connectivity component of γ^r and there is no circuit in γ^r which goes through $\hat{v}^r$, l_0 and some vertex $v_i \in V^k \setminus V^r$, and $y^r_{l_0}(\gamma) = 0$ elsewhere. As a result we come to

$$Y^r_{l_0}(\boldsymbol{p}, \boldsymbol{t}, \boldsymbol{\tau}) = \prod_l t_l^{\mathfrak{N}(\hat{\gamma}_l^r)+y^r_{l_0}(\gamma_l)} \prod_{l'} \tau_{l'}^{\mathfrak{M}^{\hat{\Gamma}^r}(\gamma'_{l'})+y^r_{l_0}(\gamma_{l'-1})} P^{r,Y}_{l_0}(\boldsymbol{t}, \boldsymbol{\tau}). \tag{A.9}$$

Finally, turn to the factorization of the forms $K^r_{l_1l_2}$ which enter to the quadratic terms of the exponent. Definition (4.7) shows that $K^r_{l_1l_2}$ is the sum $\sum_T (\pm\Pi_T)$ over pseudo-trees of the graph $\hat{\Gamma}^r$ with a circuit going through l_1 and l_2 (consider separately the case of diagonal and non-diagonal elements). It is easy to see that $-\mathfrak{N}(\gamma^r \cap T)$ takes the minimal value $-\chi^r_{l_1l_2}(\gamma)$ and $\chi^r_{l_1l_2}(\gamma) = 1$ if there is a circuit in γ^r which goes through $l_{1,2}$, and $\chi^r_{l_1l_2}(\gamma) = 0$ elsewhere. By taking into account the non-negativity of the defect we obtain

$$K^r_{l_1l_2}(t, \boldsymbol{\tau}) = \prod_l t_l^{\mathfrak{N}(\hat{\gamma}_l^r)-\chi^r_{l_1l_2}(\gamma_l)} \prod_{l'} \tau_{l'}^{\mathfrak{M}^{\hat{\Gamma}^r}(\gamma'_{l'})+1-\chi^r_{l_1l_2}(\gamma_{l'-1})} P^r_{l_1l_2}(\boldsymbol{t}, \boldsymbol{\tau}). \tag{A.10}$$

The factorization of the diagonal terms $K^r_{l_1l_1}$ turns to be "exact", as in the case of the form A^r. In other words, the polynomial $P^r_{l_1l_1}(\boldsymbol{t}, \boldsymbol{\tau})$ in (A.10) looks like $1 + \bar{P}^r_{l_1}(\boldsymbol{t}, \boldsymbol{\tau})$ with non-negative $\bar{P}^r_{l_1}$. This is rather natural because $K_{ll} = \beta_l A^{r,\{\Gamma\setminus\{l\}\}}$ where $A^{r,\{\Gamma\setminus\{l\}\}}$ is the form A^r written for the graph Γ with l-th line deleted.

Appendix B. Factorization in S-variables

To prove the factorization in S-sector variables let us use the representation

$$\Pi_T(\boldsymbol{t}', \tau') = \prod_{\gamma \in \mathcal{F}_\alpha} (t_\gamma')^{\mathfrak{N}(\hat{\gamma}^r)+\mathrm{def}_T(\hat{\gamma}^r)-\mathfrak{N}(\hat{\gamma}^r \cap T)} \prod_{\gamma \in \mathcal{F}_\beta} (\tau_\gamma')^{\mathfrak{M}^{\hat{\Gamma}^r}(\gamma)+\mathrm{def}_T(\hat{\hat{\gamma}}^r)-\mathfrak{N}(\hat{\hat{\gamma}}^r \cap T)+1-n} \tag{B.1}$$

which can be derived as (A.1). To do this, the product $\Pi_T(\boldsymbol{t},' \boldsymbol{\tau}')$ is initially written as follows

$$\prod_{\gamma \in \mathcal{F}_\alpha} (t_\gamma')^{L(\hat{\gamma}^r \cap T)} \prod_{\gamma \in \mathcal{F}_\beta} (\tau_\gamma')^{L(\hat{\gamma}^r \cap T)-r}.$$

Eventually, the factorization in S-variables for all the polylinear forms except A^r becomes a consequence of H-sector factorization that is described by formulae (A.5) to (A.10). In fact, S-sector factorization is the same apart from the changes

$$\prod_l \to \prod_{\gamma\in\mathcal{F}_\alpha}, \qquad \prod_{l'} \to \prod_{\gamma\in\mathcal{F}_\beta}, \qquad \gamma_l \to \gamma \in \mathcal{F}_\alpha, \qquad \gamma'_{l'} \to \gamma \in \mathcal{F}_\beta, \qquad \gamma_{l'-1} \to \tilde{\gamma}.$$

Let us now establish the factorization

$$A^r(\boldsymbol{t}', \boldsymbol{\tau}') = \prod_{\gamma\in\mathcal{F}_\alpha} (t_\gamma')^{\mathfrak{N}(\gamma)} \prod_{\gamma\in\mathcal{F}_\beta} (\tau_\gamma')^{\hat{\mathfrak{M}}(\gamma)} \left(1 + P_A(\boldsymbol{t}', \boldsymbol{\tau}')\right) \tag{B.2}$$

where it is understood that the mixed representation is appropriately chosen for given $\mathcal{L}_\alpha$. To prove (B.2) we first rewrite (B.1) for the form A^r

$$\prod_{\gamma\in\mathcal{F}_\alpha} (t_\gamma')^{\mathfrak{N}(\hat{\gamma}^r)+\mathrm{def}_T(\hat{\gamma}^r)} \prod_{\gamma\in\mathcal{F}_\beta} (\tau_\gamma')^{\mathfrak{M}^{\hat{\Gamma}^r}(\gamma)+\mathrm{def}_T(\hat{\tilde{\gamma}}^r)}.$$

Here $\mathfrak{N}(\hat{\gamma}^r) = \mathfrak{N}(\gamma)$ at $\mathcal{L}(\gamma) \subset \mathcal{L}_\alpha$ and $\mathfrak{M}^{\hat{\Gamma}^r}(\gamma) = \hat{\mathfrak{M}}(\gamma)$ at $\mathcal{L}(\gamma) \subset \mathcal{L}_\beta$ due to the proper choice of $\mathcal{V}^r$. Then it is necessary to find the tree T of $\hat{\Gamma}^r$ which provides the minimal defect for all $\gamma \in \mathcal{F}_\alpha$ and $\gamma \in \mathcal{F}_\beta$, i.e. $\mathrm{def}_T(\hat{\gamma}^r) = 0$ and $\mathrm{def}_T(\hat{\tilde{\gamma}}^r) = 0$, respectively.

We choose the tree as follows:

$$\mathcal{L}(T) =: \mathcal{L}^T = \mathcal{L}_\alpha{}^T \cup \mathcal{L}_\beta{}^T \qquad \text{with}$$

$$\mathcal{L}_\alpha{}^T = \{l \mid l \in \gamma \in \mathcal{F}_\alpha, \quad L(\gamma) = 1\},$$

$$\mathcal{L}_\beta{}^T = \mathcal{L}_\beta \setminus \mathcal{L}_\beta{}^{T^*}, \qquad \mathcal{L}_\beta{}^{T^*} = \{l \mid l \in \gamma \in \mathcal{F}_\beta', \quad L(\gamma) = 1\}.$$

Now it suffices to verify that T, indeed, possesses the desired property.

Before doing this we note that if $\mathcal{F}$ is a maximal forest then for any $\gamma \in \mathcal{F}_\alpha$ the set $\{\gamma' \in \mathcal{F}_\alpha \mid \gamma' \subset \gamma\} = F_\alpha{}^\gamma$ is a maximal UV forest of γ. Analogously, for any $\gamma \in \mathcal{F}_\beta$ the set $\{\gamma' \in \mathcal{F}_\beta \mid \gamma' \subset \gamma\} = \mathcal{F}_\beta{}^\gamma$ is a maximal IR forest of $\Gamma/\tilde{\gamma}$. Furthermore, $\mathcal{F}_\alpha{}^\gamma$ has exactly $\mathfrak{N}(\gamma)$ elements with $L(\gamma) \geqq 2$ (see e.g. Appendix A of ref. [13]) and $F_\beta{}^\gamma$ has exactly $\mathfrak{M}(\hat{\Gamma}/\tilde{\gamma})$ elements with $L(\gamma) \geqq 2$. Hence, the reduction of T_α to any $\gamma \in \mathcal{F}_\alpha$ is a tree of γ since it is a maximal subgraph without circuits. Therefore, $\mathrm{def}_T(\hat{\gamma}^r) = 0$. Similarly, the reduction of $\mathcal{L}_\beta{}^{T^*}$ to any $\gamma \in \mathcal{F}_\beta$ becomes a co-tree of $\hat{\Gamma}/\tilde{\gamma}$ since it is a maximal subgraph without cut-sets. Thus, its complement in $\hat{\Gamma}/\tilde{\gamma}$ becomes a tree, and, therefore, the reduction of T to every connectivity component of $\tilde{\gamma}$ is nothing but a tree. This leads to $\mathrm{def}_T(\hat{\tilde{\gamma}}^r) = 0$ for any $\gamma \in \mathcal{F}_\beta$.

References

[1] Bogoliubov, N. N., Shirkov, D. V., Introduction to the Theory of Quantized Fields (N. Y., Interscience Publ. Inc., 1959).
[2] Zavialov, O. I., Renormalized Feynman Diagrams (Moscow, Nauka, 1979; in Russian).
[3] Speer, E. R., J. Math. Phys. **9**, 1404 (1968); Generalized Feynman Amplitudes (Princ. Univ. Press, 1969).
[4] Hepp, K., Commun. Math. Phys. **2**, 301 (1966); Théorie de la Renormalisation (Berlin—Heidelberg—New York, Springer, 1969).
[5] Appelquist, T., Ann. Phys. **54**, 27 (1969).
[6] Nakanishi, N., Graph Theory and Feynman Integrals (N.Y., Gordon and Breach, 1971).
[7] Anikin, S. A., Polivanov, M. C., Zavialov, O. I., Theor. Math. Phys. **17**, 1982 (1973).
[8] Trute, H., Prepr. DESY 74/44 (1974).
[9] Cvitanović, P., Kinoshita, T., Phys. Rev. **D 10**, 3978; 3991 (1974).

[10] SPEER, E. R., Ann. Inst. H. Poincaré **23**, 1 (1975).
[11] BERGÈRE, M. C., LAM, Y.-M., J. Math. Phys. **17**, 1546 (1976); Commun. Math. Phys. **48**, 267 (1976).
[12] BERGÈRE, M. C., DAVID, F., J. Math. Phys. **20**, 1244 (1979); Commun. Math. Phys. **81**, 1 (1981).
[13] BREITENLOHNER, P., MAISON, D., Commun. Math. Phys. **52**, 39, 55 (1977).
[14] ANIKIN, S. A., POLIVANOV, M. C., ZAVIALOV O. I. Fortschr. Phys. **27**, 527 (1977); ANIKIN, S. A., ZAVIALOV, O. I., Ann. Phys. **116**, 135 (1978); ANIKIN, S. A., ZAVIALOV, O. I., Theor. Math. Phys. **27**, 585 (1976).
[15] SMIRNOV, V. A., Theor. Math. Phys. **44**, 761 (1980); **46**, 132 (1981).
[16] ANIKIN, S. A., KARCHEV, N. I., ZAVIALOV, O. I., Theor. Math. Phys. **44**, 751 (1980); **46**, 1 (1981).
[17] HAMPTON, R. R., J. Math. Phys. **22**, 1749 (1981).
[18] GEL'FAND, I. M., SHILOV, G. E., Generalized Functions (N.Y., Acad. Press, Inc., 1964).
[19] BERNSTEIN, I. N., GEL'FAND, S. I., Funct. Anal. Appl. (in Russian) **3** (1), 84 (1969); BERNSTEIN, I. N., ibid. **6** (4), 26 (1972); ATIYAH, M. F., Comm. Pure Appl. Math. **23**, 145 (1970).
[20] KOSINSKI, P., MASLANKA, P., J. Phys. A: Math. Gen. **12**, L 133, (1977).
[21] BLECHER, P. M., prepr. IAM, No. 117, 141 (1979).
[22] ZIMMERMANN, W., Ann. Phys. **77**, 536; 570 (1973).
[23] CHETYRKIN, K. G., TKACHOV, F. V., Phys. Lett. **114 B**, 340 (1982).
[24] TKACHOV, F. V., Phys. Lett. **124 B**, 212 (1983); CHETYRKIN, K. G., ibid. **126 B**, 371 (1983); CHETYRKIN, K. G., GORISHNY, S. G., LARIN, S. A., TKACHOV, F. V., ibid. **132 B**, 351 (1983).
[25] SMIRNOV, V. A., Theor. Math. Phys. **36**, 676 (1978); **41**, 1056 (1979); **46**, 17 (1981); **47**, 140 (1981).
[26] HARARY, F., Graph Theory (London, Addison-Wesley, 1969).
[27] LOWENSTEIN, J. H., SPEER, E. R., Commun. Math. Phys. **47**, 43, (1976).
[28] KIRCHHOFF, G., Ann. Phys. Chem. **72**, 497 (1847).
[29] LOWENSTEIN, J. H., ZIMMERMANN, W., Commun. Math. Phys. **44**, 73 (1975); LOWENSTEIN, J. H., ibid. **47**, 53 (1976).
[30] CHETYRKIN, K. G., SMIRNOV, V. A., Teor. Mat. Fiz. **56**, 206 (1983).
[31] DE CALAN, C., MALBOUISSON, A. P. C., Commun. Math. Phys. **90**, 413 (1983).
[32] 'T HOOFT, G., VELTMAN, M., Nucl. Phys. **B 44**, 189 (1972); BOLLINI, G. G., GIAMBIAGI, J. J., Nuovo Cim. **12 B**, 20 (1972).
[33] CHETYRKIN, K. G., TKACHOV, F. V., Nucl. Phys. **B 192**, 159 (1981).

Fortschr. Phys. 33 (1985) 9, 523—546

Renormalization and Operator Product Expansion in Theories with Massless Particles

S. A. Anikin

Institute of Chemical Physics, Academy of Science, USSR

V. A. Smirnov

Nuclear Physics Institute of Moscows State University,
Moscow, USSR

Renormalization procedure in theories including massless particles is presented. With the help of counterterm formalizm the operator product expansion for arbitrary composite fields is derived. The coefficient functions are explicitly expressed in terms of certain Green's functions.

Contents

1. Introduction

The purpose of this paper is to describe two kinds of renormalization procedure for theories including massless particles. The most part of the paper deals with a generalization of the soft-mass renormalization. We prove the Bogoliubov-Parasiuk theorem and develop counterterm formalism (CTF) which provides a definition of composite fields and their products. This makes it possible to derive the operator product expansion (OPE) for arbitrary composite fields. Furthermore, the BP-theorem is also demonstrated for modified analytic renormalization.

As is well known the standard BPHZ renormalization [1—4], generally, fails in theories including massless particles because of the infrared (IR) divergences involved. Different modifications applicable to these theories have been proposed [5—7]. Any procedure of this type differs from the standard one by the "last" subtractions which are performed with an IR regularization. In the case of minimal subtractions of this type IR, as well as ultraviolet (UV), divergences are removed from a wide class of theories denoted in ref. [7] as "theories almost without IR divergences". Note that for these theories the BP-theorem was also demonstrated in the dimensional [8] and, for massless case, in the analytic renormalization [9].

In should be emphasized that the BPHZ renormalization is now the only renormalization procedure making it possible to derive the OPE explicitly. This expansion proposed by Wilson [10] and first proved in the simplest case by Zimmermann [11] looks like

$$\varphi_1(x + \varrho\xi_1)\, \varphi_2(x + \varrho\xi_2)^{\varrho\to 0} \sim \sum_{i=1}^{n} K_i(\xi_1, \xi_2, \varrho)\, J^i(x) + Q^{N+1}(x, \varrho\xi_1, \varrho\xi_2) \qquad (1.1)$$

where $\xi^2 \neq 0$, $\xi = \xi_1 - \xi_2$; $J^i(x)$ are composite fields (currents); $K_i(\xi_1, \xi_2, \varrho)$ are coefficient functions exhibiting the singularities at $\xi = 0$ and being explicitly expressed in terms of some Green's functions. The expansion is asymptotic: for any N there is n such that the remainder has the order ϱ^N.

By means of the CTF [12] the result was then generalized for a product of arbitrary composite fields [13, 4]. This formalism considerably simplifies derivations as all the combinatorics of the R-operation is resolved explicitly. Note that in order to derive the OPE it was necessary to make oversubtractions in diagrams including vertices $x_{1,2} = x + \varrho\xi_{1,2}$ simultaneously.

Our paper may be regarded as a generalization of ref. [4] for theories including massless particles. On the other hand, the proofs of analytical results involved are essentially based on the method developed in ref. [15] and presented, in extended form, in the preceding paper.

Let us now describe a renormalization procedure applicable to theories with massless particles.

2. Renormalization of theories including massless particles

1. In perturbative quantum field theory with massless particles additional ambiguities in Green's functions, generally, appear due to IR divergences. As for UV divergences, they are removed by the well-known renormalization technique. As a result, it is possible to make a theory UV-finite by inserting local counterterms into Lagrangian. If a finite number of counterterms with different operator structure is needed then the theory is called renormalizable. In particular, all the theories with dimensionless coupling constants belong to this class.

IR divergences are, in general, removed by non-local counterterms. A class of theories in which IR divergences are eliminated by standard local counterterms was described in ref. [7]: these are "theories almost without IR divergences". In four-dimensional space-time the theories with interaction Lagrangians composed of the terms $\partial^{2k}(\bar{F}F)^{2q+1} B^{2p+1}$; $\partial^{2k+1}(\bar{F}F)^{2q} B^{2p+1}$; $\partial^k(\bar{F}F)^q B^{2p} (p > 0)$ belong to this class. Here $F(B)$ are fermion (boson) fields which may be of different types. The number k indicates the total number of derivatives in the corresponding term. At some values of (k, q, p) the terms leading to nonrenormalizability appear. However, one may consider a finite number of such vertices when dealing with a renormalizable theory. Note that the theories with dimensionless coupling constants are almost without IR divergences. In ref. [7] it was demonstrated that in a theory almost without IR divergences these divergences may appear only due to self-energy insertions in massless lines.

Our first purpose is to define a renormalization procedure that removes both UV and IR divergences with the help of local counterterms. The necessary physical restrictions are automatically satisfied if the renormalization is accomplished by means of R-operation. We believe that the standard R-operation (see e.g. ref. [4]), as well as the α-representation of Feynmann amplitudes (FA), are well-known. In what follows we present two different renormalization schemes. The main results concern a renormalization of BPHZ-type. Furthermore, in Sect. 6 a modification of the analytic renormalization [14] is described.

2. BPHZ renormalization is based on a subtracting operator that acts in the momentum space. To define this operator let us consider a diagram γ and the corresponding FA written in the α-representation

$$G_\gamma(\boldsymbol{p}) = \int d\boldsymbol{\alpha} G_\gamma(\boldsymbol{p}, \boldsymbol{\alpha}). \tag{2.1}$$

Here $\boldsymbol{p} = (p_1, \ldots, p_n)$ are external momenta of γ, $\boldsymbol{\alpha} = (\alpha_1, \ldots, \alpha_L)$, $d\boldsymbol{\alpha} = d\alpha_1 \ldots d\alpha_L$, and α_l are parameters associated with L lines of γ.

Now we modify each massless propagator by inserting into its α-representation an additional factor $\varphi(\alpha\mu^2)$ which performs IR regularization. An admissible class of functions φ will be described in the end of Sect. 5.1. Here μ is a massive parameter. In the case of "soft-mass" renormalization μ is nothing but the soft mass. Different propagators may be regularized independently, however, to provide a self-consistent procedure it is necessary to introduce the same function φ into the propagators of the field of a given type. Note that massive propagators may be also IR regularized. This change leads to a finite renormalization which is always involved in any R-operation. Such complementary regularization becomes useful when analyzing the asymptotic behaviour of Green's functions in the massless limit.

The procedure described provides a modification of the function $G_\gamma(\boldsymbol{p}, \boldsymbol{\alpha})$ in (2.1) which transforms into $G_\gamma(\boldsymbol{p}, \boldsymbol{\alpha}, \mu^2)$. Let us now correspond two numbers a_γ, b_γ with $b_\gamma \leqq a_\gamma$ to any (sub)graph γ. The subtracting operator is defined as follows

$$M_\gamma^{a_\gamma, b_\gamma} = M_\gamma^{b_\gamma} I_\gamma + \sum_{k=b_\gamma+1}^{a_\gamma} \Delta_\gamma^{a_\gamma-k}(M_\gamma^k - M_\gamma^{k-1}) \tag{2.2}$$

where

$$I_\gamma G_\gamma(\boldsymbol{p}, \boldsymbol{\alpha}) = G_\gamma(\boldsymbol{p}, \boldsymbol{\alpha}, \mu^2 - \zeta^2\mu^2)|_{\zeta=1},$$

$$\Delta_\gamma^n G_\gamma(\boldsymbol{p}, \boldsymbol{\alpha}) - \mathcal{M}_\zeta^n G_\gamma(\boldsymbol{p}, \boldsymbol{\alpha}, \mu^2 \quad \zeta^2\mu^2), \tag{2.3}$$

$$M_\gamma^n G_\gamma(\boldsymbol{p}, \boldsymbol{\alpha}, \mu^2) = \mathcal{M}_\varkappa^n G_\gamma(\varkappa\boldsymbol{p}, \boldsymbol{\alpha}, \mu^2), \tag{2.4}$$

$$\mathcal{M}_\sigma^n f(\sigma) = \sum_{k=0}^{n} \mathcal{M}_\sigma^{(n)} f(\sigma) = \sum_{k=0}^{n} \frac{1}{k!} f^{(k)}(0). \tag{2.5}$$

Note that, in accordance with (2.3), the operators I_γ and Δ_γ^n act on the factor $\prod_{l\in\gamma} \varphi(\alpha_l\mu^2)$ only. If a line l belong to subdiagrams $\gamma_1, \ldots, \gamma_n$ which are operated either by $\Delta_{\gamma_j}^{n_j}$ or by I_{γ_i} then the corresponding parameters ζ_j enter as follows

$$\varphi(\alpha_l\mu^2) \to \varphi\left(\alpha_l\left(\mu^2 - \prod_{j=1}^{n} \zeta_j^2\mu^2\right)\right).$$

The operator $M_\gamma^{a_\gamma, b_\gamma}$ performs the Taylor expansion of FA $G_\gamma(\boldsymbol{p})$ in the external momenta up to degree b_γ, and the next terms of the expansion (up to degree a_γ) acquire the IR regularization due to functions φ. Let us refer to these terms as to "massive" ones because the functions φ and their derivatives behave like massive factors in the propagators.

Now we introduce the R-operation by means of the standard formula

$$R = \sum_{F_0} \prod_{\gamma\in F_0} (-M_\gamma^{a_\gamma, b_\gamma}) \tag{2.6}$$

where the sum goes over the forests F_0 of generalized vertices, i.e. complete 1 PI subgraphs, of a given graph. As usual a forest is a set $\{\gamma_i\}$ of non-overlapping subgraphs (i.e. $\gamma_i \cap \gamma_j = \emptyset$, or $\gamma_i \subset \gamma_j$, or $\gamma_i \supset \gamma_j$ for any pair of these γ). In the case of renormalization (2.6) the Bogoliubov-Parasiuk theorem takes place.

Theorem 2.1. The renormalized FA $G_\Gamma^R(p) \equiv RG_\Gamma(p)$ regarded as a tempered distribution is UV- ahd IR-finite if the following conditions are fulfilled by the numbers a_γ and b_γ

A) $0 \leqq a_\gamma - \omega_\gamma \equiv a^\gamma; \qquad 0 \leqq \omega_\gamma - 1 - b_\gamma \equiv b^\gamma.$

B) For any set $\{\gamma_i\}$ of pairwise disjoint subgraphs with $\gamma_i \Subset \gamma$ the inequalities

$$\sum_i a^{\gamma_i} \leqq a^\gamma; \qquad \sum_i b^{\gamma_i} \leqq b^\gamma$$

hold.

A proof of the theorem will be presented in Sect. 5.

3. Counterterm formalism

1. In this section we shall recall the main relations of the CTF and adapt them for the theories including massless particles. Let us first introduce the notations which will be of current use. The normal product of asymptotic fields considered at the same point will be denoted by $b_{\{\lambda\}}(x) = :\varphi_i^{(\lambda_1)}(x) \ldots \varphi_l^{(\lambda_l)}(x):$. Here $\{\lambda\} = \{\lambda_1, \ldots, \lambda_l\}$,

$$\lambda_j = (\lambda_{j0}, \lambda_{j1}, \lambda_{j2}, \lambda_{j3}), \qquad \varphi^{(\lambda_j)}(x) = \left(\frac{\partial}{\partial x_0}\right)^{\lambda_{j0}} \cdots \left(\frac{\partial}{\partial x_3}\right)^{\lambda_{j3}} \varphi(x).$$

The dimension of $b_{\{\lambda\}}(x)$ in mass units is

$$d_{\{\lambda\}} \equiv \dim b_{\{\lambda\}}(x) = \sum_{j=1}^{l} (\dim \varphi_j + |\lambda_j|)$$

with $|\lambda_j| = \lambda_{j0} + \lambda_{j1} + \lambda_{j2} + \lambda_{j3}$. Furthermore, the notations $\{\lambda\}! = \prod_{j=1}^{l} (\lambda_j)!, = (\lambda_j)! = \lambda_{j0}! \ldots \lambda_{j3}!$ will be used. Let us denote by $b_{[k]}(y_1, \ldots, y_k) \equiv b_{[k]}(\boldsymbol{y})$ the product of fields (without derivatives) taken at different points:

$$b_{[k]}(\boldsymbol{y}) = :\varphi_1(y_1) \ldots \varphi_k(y_k):, \qquad d_{[k]} \equiv \dim b_{[k]}(y) = \sum_{j=1}^{k} \dim \varphi_j.$$

If there are $k_1, \ldots, k_N$ fields of N different types in $b_{[k]}$ then $[k]! = k_1! \ldots k_N!$ with $k_1 + \cdots + k_N = k$. Finally,

$$\tilde{b}^{\{\lambda\}}(0) = \left(\frac{\partial}{\partial q_1}\right)^{(\lambda_1)} \cdots \left(\frac{\partial}{\partial q_l}\right)^{(\lambda_l)} \tilde{\varphi}_1(q_1) \ldots \tilde{\varphi}_l(q_l)|_{q=0}.$$

Note that all internal indices of the fields φ_j are included, if necessary, into $[k]$ and $\{\lambda\}$.

Any operator in Fock space is written in terms of its normal symbol off the mass shell and is represented according to the expansion in asymptotic fields

$$F(x_1, \ldots, x_n) = \sum_{[k]} \frac{1}{[k]!} \int F_{[k]}(\boldsymbol{x} \mid y_1, \ldots, y_k)\, b_{[k]}(\boldsymbol{y})\, d\boldsymbol{y}. \tag{3.1}$$

The sum runs over all possible $b_{[k]}$. The coefficient functions $F_{[k]}(\boldsymbol{x} \mid \boldsymbol{y})$ are supposed to be expanded in FA in framework of the perturbation theory.

2. In Sect. 2 the renormalization of diagrams was discussed. In ref. [4] it was described how to turn to the renormalization of the perturbation theory series as a whole. To do

this, one should define subtracting operator acting on a functional of the type (3.1) rather than on a diagram. In ref. [4] the operator M^a was introduced as follows

$$M^a F(x_1, \dots, x_n) = \sum_{\{\lambda\}:d_{\{\lambda\}}\leqq a} c_{\{\lambda\}} \int dx b_{\{\lambda\}}(x) \times \left\{ \sum_{\{\nu\}:\Sigma|\nu|+d_{\{\lambda\}}\leqq a} \frac{(-i)^{\Sigma|\nu_j|}}{\{\nu\}!} \delta^{(\nu_1)}(x - x_1) \dots \delta^{(\nu_n)}(x - x_n) \, \Phi^{\{\lambda\}}_{\{\nu\}}[F] \right\} \tag{3.2}$$

with $c_{\{\lambda\}} = (-i)^{\Sigma|\lambda_j|} ([l]! \, \{\lambda\}!)^{-1}$,

$$\Phi^{\{\lambda\}}_{\{\nu\}}[F] = \prod_{i=1}^{n} \left(\frac{\partial}{\partial p_i}\right)^{(\nu_i)} \prod_{j=1}^{l} \left(\frac{\partial}{\partial k_j}\right)^{(\lambda_j)} \tilde{F}^{\text{prop}}_{[l]}(\boldsymbol{p} \mid \boldsymbol{k})|_{\boldsymbol{p}=\boldsymbol{k}=0},$$

$$(2\pi)^4 \, \delta(\Sigma p + \Sigma k) \, \tilde{F}^{\text{prop}}_{[l]}(p \mid k) = \int d\boldsymbol{x} \, d\boldsymbol{y} e^{i\Sigma p\boldsymbol{x} + i\Sigma k\boldsymbol{y}} F^{\text{prop}}_{[l]}(\boldsymbol{x} \mid \boldsymbol{y}).$$

The symbol "prop" means the restriction to the contribution of amputated 1PI diagrams. In the case of $n = 1$ formula (3.2) is simplified

$$M^a F(x) = \sum_{\{\lambda\}:d_{\{\lambda\}}\leqq a} b_{\{\lambda\}}(x) \, \Phi^{\{\lambda\}}[F] \tag{3.3}$$

where $\Phi^{\{\lambda\}}[F] = \langle F(0) \, \tilde{b}^{\{\lambda\}}(0)\rangle^{\text{prop}} \, c_{\{\lambda\}}$.

The CTF in theories without IR divergences is based on the M^a that is the "functional" extension of the operator $M_\gamma{}^a$ introduced by means of (2.4). But if IR divergences are involved an extension of $M_\gamma{}^{a_\gamma, b_\gamma}$ should be used. For its definition two auxiliary operators I and Δ^n are necessary

$$\begin{aligned} I F_{[k]}(\boldsymbol{x} \mid \boldsymbol{y}) &= F_{[k]}(\boldsymbol{x} \, |\boldsymbol{y}| \, \mu^2 - \zeta^2\mu^2)|_{\zeta=1}, \\ \Delta^n F_{[k]}(\boldsymbol{x} \mid \boldsymbol{y}) &= \mathcal{M}_\zeta{}^n F_{[k]}(\boldsymbol{x} \, |\boldsymbol{y}| \, \mu^2 - \zeta^2\mu^2). \end{aligned} \tag{3.4}$$

Here $F_{[k]}(\boldsymbol{x} \, |\boldsymbol{y}| \, \mu^2)$ is FA regularized with the help of the function φ (see Sect. 2). Although the operator I simply removes the IR regularization it acts nontrivially when several operators I and Δ^n are involved simultaneously. Now we define the operator $M^{a,b}$ as follows

$$M^{a,b} = \begin{cases} M^a I, & a \leqq b; \\ M^b I + \sum\limits_{k=b+1}^{a} \Delta^{a-k}(M^k - M^{k-1}), & a > b. \end{cases} \tag{3.5}$$

3. The operator $M^{a,b}$ makes it possible to adapt all the formulae of CTF [4] for theories including massless particles. For simplicity, only theories with dimensionless coupling constants will be considered. First, we have to introduce the S-matrix $S = RTe^s$ where $s = i \int \mathcal{L}(x) \, dx$ is the unrenormalized action corresponding to the Lagrangian $\mathcal{L}(x)$. Note that all the products involved will be understood as time-ordered products. If the ordinary product is needed we shall use the symbol $\otimes$. Furthermore, the following simple notations will be used

$$E_0(s) = T \exp s; \qquad E_1(s) = -1 + T \exp s;$$

$$E_2(s) = -1 - s + T \exp s.$$

The combinatorics of Lemma 5.1 of ref. [4] leads straightforwardly to the same result

$$RE_0(s) \equiv E_0(s_r) = E_0\big(s - ME_2(s - ME_2(s - \dots))\big).$$

The only difference is that when diagrammatic expansion is considered the operator M transforms into $M_\gamma^{\omega_\gamma,\omega_\gamma-1}$ which performs the minimal renormalization.

Now we introduce the composite field $B^{a,b}_{\{\lambda\}}(x) = R^{a,b}E_0(s)\, b_{\{\lambda\}}(x)$. The indices a, b correspond to the vertex x and specify the renormalization of diagrams. As in Lemma 5.2 of ref. [4] we have

$$B^{a,b}_{\{\lambda\}}(x) \equiv E_0(s_r)\, 'B^{a,b}_{\{\lambda\}}(x) = E_0(s_r) \frac{1}{1 + M^{a,b}E_1(s_r)}\, b_{\{\lambda\}}(x). \tag{3.6}$$

Note that

$$M^{a,b}B^{a,b}_{\{\lambda\}}(x) = b_{\{\lambda\}}(x). \tag{3.7}$$

Finally, we introduce the product of two composite fields

$$\left(B^{a_1,b_1}_{\{\mu\}}(x_1)\, B^{a_2,b_2}_{\{\nu\}}(x_2)\right)^{a,b}_+ = RE_0(s)\, b_{\{\mu\}}(x_1)\, b_{\{\nu\}}(x_2).$$

In accordance with Lemma 5.3 of ref. [4] we get

$$\left(B^{a_1,b_1}_{\{\mu\}}(x_1)\, B^{a_2,b_2}_{\{\nu\}}(x_2)\right)^{a,b}_+ = E_0(s_r) \frac{1}{1 + M^{a,b}E_1(s_r)} (1 - M^{a,b})\, 'B^{a_1,b_1}_{\{\mu\}}(x_1)\, 'B^{a_2,b_2}_{\{\nu\}}(x_2). \tag{3.8}$$

The lhs of this expression is well-defined if $a \geqq a_1 + a_2 - 4$: this result follows directly from Theorem 2.1.

4. The Zimmermann identities hold for the composite fields $B^{a,b}_{\{\lambda\}}(x)$ as well as for $B^a_{\{\lambda\}}(x)$. To derive these identities let us take into account that $M^{a,b}M^{a',b'} = M^{a',b'}$ if $a' \leqq a$, $b' \leqq b$. Then one easily verifies the following relation

$$B^{a',b'}(x) = B^{a,b}(x) + E_0(s_r) \frac{1}{1 + M^{a,b}E_1(s_r)} (M^{a,b} - M^{a',b'})\, B^{a',b'}(x). \tag{3.9}$$

Here non-essential indices $\{\lambda\}$ are omited. Now we use definition (3.5). By setting $\Delta^n = 0$ at $n < 0$ we come to

$$M^{a,b} - M^{a',b'} = \sum_{k=b'+1}^{b} (I - \Delta^{a'-k})\, (M^k - M^{k-1}) + \sum_{k=b+1}^{a} (\Delta^{a-k} - \Delta^{a'-k})\, (M^k - M^{k-1}). \tag{3.10}$$

Inserting (3.10) into (3.9) and applying (3.3) yields the following result

$$B^{a,b}(x) - B^{a',b'}(x) = E_0(s_r)\, [1 + M^{a,b}E_1(s_r)]^{-1} \times \Bigg\{ \sum_{k=b'+1}^{b} \sum_{\{\lambda\};d_{\{\lambda\}}=k} (I - \Delta^{a'-k})\, \Phi^{\{\lambda\}}[B^{a',b'}]\, b_{\{\lambda\}}(x) + \sum_{k=b+1}^{a} (\Delta^{a-k} - \Delta^{a'-k}) \sum_{\{\lambda\}:d_{\{\lambda\}}=k} \Phi^{\{\lambda\}}[B^{a',b'}]\, b_{\{\lambda\}}(x) \Bigg\}. \tag{3.11}$$

Then we represent Δ^n as $\sum\limits_{k=0}^{n} (\Delta^k - \Delta^{k-1}) \equiv \sum\limits_{k=0}^{n} \Delta_k$ and obtain

$$\Delta^{a-k} - \Delta^{a'-k} = \sum_{n=a'+1}^{a} \Delta_{n-k}. \tag{3.12}$$

Furthermore, we note that for any functional F_1 the following relations take place

$$M^{a,b}\{F_1 I \Phi^{\{\lambda\}}[B]\} = \Phi^{\{\lambda\}}[B]\, M^{a,b} F_1,$$

$$M^{a,b}\{F_1 \Delta_{a-n} \Phi^{\{\lambda\}}[B]\} = \{\Delta_{a-n} \Phi^{\{\lambda\}}[B]\}\, M^{n,b} F_1.$$

Finally, we have the desired statement.

Lemma 3.1. If $a \geqq a' \geqq b'$, $a \geqq b \geqq b'$ then

$$B^{a',b'}_{\{\mu\}}(x) = B^{a,b}_{\{\mu\}}(x) + \sum_{k=b'+1}^{b} \sum_{\{\lambda\}:d_{\{\lambda\}}=k} \left\{\Phi^{\{\lambda\}}[B^{a',b'}_{\{\mu\}}]\, B^{a,b}_{\{\lambda\}}(x) - \sum_{n=k}^{a'} \{\Delta_{n-k} \Phi^{\{\lambda\}}[B^{a',b'}_{\{\mu\}}]\}\, B^{a-n+k,b}_{\{\lambda\}}(x)\right\}$$
$$+ \sum_{k=b+1}^{a} \sum_{\{\lambda\}:d_{\{\lambda\}}=k} \sum_{n=a'+1}^{a} \{\Delta_{n-k} \Phi^{\{\lambda\}}[B^{a',b'}_{\{\mu\}}]\}\, B^{a-n-k,b}_{\{\lambda\}}(x). \tag{3.13}$$

Corollary 1. If $b' = b$, $a' = a - 1$ then

$$B^{a-1,b}_{\{\mu\}}(x) = B^{a,b}_{\{\mu\}}(x) + \sum_{k=b+1}^{a} \sum_{\{\lambda\}:d_{\{\lambda\}}=k} \{\Delta_{a-k} \Phi^{\{\lambda\}}[B^{a-1,b}_{\{\mu\}}]\}\, B^{k,b}_{\{\lambda\}}(x). \tag{3.14}$$

Corollary 2. If $b' = b - 1$, $a' = a$ then

$$B^{a,b-1}_{\{\mu\}}(x) = B^{a,b}_{\{\mu\}}(x) + \sum_{\{\lambda\}:d_{\{\lambda\}}=b} \left\{\Phi^{\{\lambda\}}[B^{a,b-1}_{\{\mu\}}]\, B^{a,b}_{\{\lambda\}}(x) - \sum_{n=b}^{a} \{\Delta_{n-b} \Phi^{\{\lambda\}}[B^{a,b-1}_{\{\mu\}}]\}\, B^{a-n+b,b}_{\{\lambda\}}(x).\right. \tag{3.15}$$

Let us now consider the product of two composite fields $\left(B^{a_1,b_1}_{\{\mu\}}(x_1)\, B^{a_2,b_2}_{\{\nu\}}(x_2)\right)^{a,b}_{+} \equiv (B_1, B_2)^{a,b}_{+}$. Apply then formulae (3.8) and divide the rhs into two parts

$$(B_1 B_2)^{a,b}_{+} = E_0(s_r)\,[1 + M^{a,b} E_1(s_r)]^{-1}\,(1 - M^{a,b})\; 'B_1 ' B_2$$
$$= (B_1 B_2)^{a,b}_{\text{loc}} + E_0(s_r)\; 'B_1 ' B_2,$$

where

$$(B_1 B_2)^{a,b}_{\text{loc}} = -E_0(s_r)\,[1 + M^{a,b} E_1(s_r)]^{-1}\, M^{a,b} E_0(s_r)\; 'B_1 ' B_2. \tag{3.16}$$

Note that in accordance with definitions (3.2) and (3.5) the term $(B_1 B_2)^{a,b}_{\text{loc}}$ takes the form

$$(B_1 B_2)^{a,b}_{\text{loc}} = \sum_{\{\lambda\}:d_{\{\lambda\}} \leqq b} \int K^{\{\lambda\}}(x - x_1, x - x_2)\, B^{a,b}_{\{\lambda\}}(x)\, dx$$
$$+ \sum_{k=b+1} \sum_{n=k} \sum_{\{\lambda\}:d_{\{\lambda\}}=k} \int \{\Delta_{n-k} K^{\{\lambda\}}(x - x_1, x - x_2)\}\, B^{a-n+k,b}_{\{\lambda\}}(x)\, dx \tag{3.17}$$

where the functions $K^{\{\lambda\}}(x - x_1, x - x_2)$ are defined by (3.2). They are linear combinations of $\delta(x - x_1)\, \delta(x - x_2)$ and its derivatives.

Let $\overline{M}^{A_j}$ $(j = 0, 1, \ldots)$ be a sequence of some operators. Then it is rather straightforward to come to the following statement that will be applied in Sect. 4 when deriving short-distance expansions.

Lemma 3.2. For any sequence of $\overline{M}^{A_j}$, the relation

$$\begin{aligned}
&\left(B^{a_1,b_1}_{\{\mu\}}(x_1)\, B^{a_2,b_2}_{\{\nu\}}(x_2)\right)^{a,b}_{+} \\
&= \left(B^{a_1,b_1}_{\{\mu\}}(x_1)\, B^{a_2,b_2}_{\{\nu\}}(x_2)\right)^{a,b}_{\mathrm{loc}} + E_0(s_r)\,\frac{1}{1+\overline{M}^{A_0}E_1(s_r)}\,\overline{M}^{A_0}E_0(s_r)\,{}'B^{a_1,b_1}_{\{\mu\}}(x_1)\,{}'B^{a_2,b_2}_{\{\nu\}}(x_2) \\
&\quad + \sum_{n=1}^{a} E_0(s_r)\,\frac{1}{1+\overline{M}^{A_n}E_1(s_r)}\left(\overline{M}^{A_n}-\overline{M}^{A_{n-1}}\right) Q^{(n-1)}_{\{\mu\}\{\nu\}}(x_1,x_2) + Q^{(a)}_{\{\mu\}\{\nu\}}(x_1,x_2)
\end{aligned} \tag{3.18}$$

holds with

$$Q^{(n)}_{\{\mu\}\{\nu\}}(x_1,x_2) = E_0(s_r)\,\frac{1}{1+\overline{M}^{A_n}E_1(s_r)}\,(1-\overline{M}^{A_n})\,{}'B^{a_1,b_1}_{\{\mu\}}(x_1)\,{}'B^{a_2,b_2}_{\{\nu\}}(x_2). \tag{3.19}$$

5. The fields of the type $B(x) = R(Sb(x))$ were understood above as composite fields However, Heisenberg fields look like $\check{B}(x) = (RS)^\dagger \otimes R(Sb(x))$ according to Gell-Mann-Low formula. Their time-ordered product is equal to

$$\begin{aligned}
\check{B}_1(x_1)\ldots\check{B}_n(x_n) &= (RS)^\dagger \otimes R\big(b_1(x_1)\ldots b_n(x_n)\,S\big) \\
&= (RS)^\dagger \otimes \big(B_1(x_1)\ldots B_n(x_n)\big)_+.
\end{aligned} \tag{3.20}$$

Since $\otimes$-product does not introduce divergences into a theory it is easy to rewrite all the results for "true" composite fields.

4. Short-distance expansion

1. With the help of the CTF developed in Sect. 3 let us now derive the OPE for two arbitrary composite fields. For the derivation of the OPE in massive theories the operator $\mathfrak{M}_x{}^a$ was used [4]

$$\mathfrak{M}_x{}^a F(x+\xi_1,\ldots,x+\xi_n) = \sum_{\{\lambda\}:d_{\{\lambda\}}\le a} b_{\{\lambda\}}(x)\, K^{\{\lambda\}}(\xi_1,\ldots,\xi_n) \tag{4.1}$$

where

$$K^{\{\lambda\}}(\xi) = c_{\{\lambda\}}\,\langle F(\xi_1,\ldots,\xi_n)\,\tilde{b}^{\{\lambda\}}(0)\rangle^{\xi\text{-prop}}. \tag{4.2}$$

The symbol "ξ-prop" implies restriction to the set of diagrams which become proper (i.e. 1 PI) after contraction of the vertices $\xi_1,\ldots,\xi_n$. The operator $\mathfrak{M}_x{}^a$ is closely related with M^a (compare (4.1) with (3.3)). Moreover,

$$\lim_{\xi\to 0} \mathfrak{M}_x{}^a F(x+\xi_1,\ldots,x+\xi_n) = M^a F(x,\ldots,x). \tag{4.3}$$

This is not surprising as $\mathfrak{M}_x{}^a$ removes the UV divergences that would appear in $\xi\to 0$. In order to derive the OPE in massless theories it is necessary to substitute $\mathfrak{M}_x{}^a$ by

$$\mathfrak{M}_x{}^{a,b} = \begin{cases} \mathfrak{M}_x{}^a I, & a\le b; \\ \mathfrak{M}_x{}^b I + \sum\limits_{k=b+1}^{a} \Delta^{a-k}(\mathfrak{M}_x{}^k - M_x{}^{k-1}), & a > b. \end{cases} \tag{4.4}$$

with the operators I and Δ^n introduced by (3.4).

Now we consider the product $\big(B_1(x+\xi_1)\,B_2(x+\xi_2)\big)^{a,b}_+$ which equals $E_0(s_r)$ $\times{}'(B_1x+\xi_1)\,{}'B_2(x+\xi_2)$ at $\xi_1 \neq \xi_2$ because the local terms vanish. Then we apply (3.18) with $\overline{M}^{A_n} = \mathfrak{M}_x^{n,b}$ and $\xi_1 \neq \xi_2$. Note that

$$[1 + \mathfrak{M}_x^{0,b} E_1(s_r)]^{-1} \mathfrak{M}_x^{0,b} E_0(s_r) \, 'B_1(x + \xi_1) \, 'B_2(x + \xi_2)$$
$$= [1 + \mathfrak{M}_x^{0} E_1(s_r)]^{-1} \langle E_0(s_r) \, 'B_1(\xi_1) \, 'B_2(\xi_2)\rangle^{\xi\text{-prop}}$$
$$= \langle E_0(s_r) \, 'B_1(\xi_1) \, 'B_2(\xi_2)\rangle^{\xi\text{-prop}}. \tag{4.5}$$

The first equality in (4.5) follows directly from the definition of $\mathfrak{M}_x^{0,b}$. To verify the second equality notice that ξ-proper diagrams do not appear here because the vertices ξ_1 and ξ_2 enter to a vacuum-type graph which can not be connected with diagrams from the expansion of $E_1(s_r)$. On the other hand, the operator $\mathfrak{M}_x^0$ nullifies every diagram that is not ξ-proper. Thus, we may rewrite (3.18) as follows

$$E_0(s_r) \, 'B_1(x + \xi_1) \, 'B_2(x + \xi_2)$$
$$= E_0(s_r) \langle 'B_1(\xi_1) \, 'B_2(\xi_2) \, E_0(s_r)\rangle^{\xi\text{-prop}}$$
$$+ \sum_{n=1}^{N} E_0(s_r) \frac{1}{1 + \mathfrak{M}_x^{n,b} E_1(s_r)} (\mathfrak{M}_x^{n,b} - \mathfrak{M}_x^{n-1,b}) \, Q^{(n-1)}(x, \xi_1, \xi_2) + Q^{(N)}(x, \xi_1, \xi_2) \tag{4.6}$$

where

$$Q^{(N)}(x, \xi_1, \xi_2) = E_0(s_r) \frac{1}{1 + \mathfrak{M}_x^{N,b} E_1(s_r)} (1 - \mathfrak{M}_x^{N,b}) \, 'B_1(x + \xi_1) \, 'B_2(x + \xi_2) \tag{4.7}$$

is the remainder of N-th order.

This is the desired expansion. To transform it to canonical form (1.1) it is necessary to rearrange the second term. The manipulations needed are the same as in the derivation of Zimmermann identities for the simplest case $b' = b$, $a' = a - 1$. Hence, we only describe the result.

Lemma 4.1. The following identity holds

$$E_0(s_r) \frac{1}{1 + \mathfrak{M}_x^{n,b} E_1(s_r)} (\mathfrak{M}_x^{n,b} - \mathfrak{M}_x^{n-1,b}) \, Q^{(n-1)}(x, \xi_1, \xi_2)$$
$$= \begin{cases} \sum\limits_{\{\lambda\}: d_{\{\lambda\}} = n} K^{\{\lambda\},n}(\xi_1, \xi_2) \, B_{\{\lambda\}}^{d_{\{\lambda\}},b}(x), & n \leqq b, \\ \sum\limits_{\{\lambda\}: b < d_{\{\lambda\}} \leqq n} K^{\{\lambda\},n}(\xi_1, \xi_2) \, B_{\{\lambda\}}^{d_{\{\lambda\}},b}(x), & n > b, \end{cases} \tag{4.8}$$

where $K^{\{\lambda\},n}(\xi_1, \xi_2) = c_{\{\lambda\}} \langle Q^{n-1}(0, \xi_1, \xi_2) \, \tilde{b}^{\{\lambda\}}(0)\rangle^{\xi\text{-prop}}$ at $n \leqq b$ and

$$K^{\{\lambda\},n}(\xi_1, \xi_2) = c_{\{\lambda\}} \Delta_{n - d_{\{\lambda\}}} \langle Q^{(n-1)}(0, \xi_1, \xi_2) \, \tilde{b}^{\{\lambda\}}(0)\rangle^{\xi\text{-prop}} \tag{4.9}$$

at $n > b$.

Substituting (4.8) into (4.7) yields the OPE.

Theorem 4.1. For arbitrary composite fields the OPE has the form

$$\left(B_{\{\mu\}}^{a_1,b_1}(x + \xi_1) \, B_{\{\nu\}}^{a_2,b_2}(x + \xi_2)\right)_+^{a,b}$$
$$= \left(B_{\{\mu\}}^{a_1,b_1}(x + \xi_1) \, B_{\{\nu\}}^{a_2,b_2}(x + \xi_2)\right)_{\text{loc}}^{a,b} + E_0(s_r) \langle E_0(s_r) \, 'B_{\{\mu\}}^{a_1,b_1}(\xi_1) \, 'B_{\{\nu\}}^{a_2,b_2}(\xi_2)\rangle^{\xi\text{-prop}}$$
$$+ \sum_{k=1}^{N} \sum_{\{\lambda\}: d_{\{\lambda\}} = k} G_{\{\mu\}\{\nu\}}^{\{\lambda\}}(\xi_1, \xi_2) \, B_{\{\lambda\}}^{d_{\{\lambda\}},b}(x) + Q_{\{\mu\}\{\nu\}}^{(N)}(x, \xi_1, \xi_2)$$

with

$$G^{\{\lambda\}}_{\{\mu\}\{\nu\}}(\xi_1, \xi_2) = \begin{cases} K^{\{\lambda\},k}_{\{\mu\}\{\nu\}}(\xi_1, \xi_2), & d_{\{\lambda\}} \leqq b; \\ \sum\limits_{k=d_{\{\lambda\}}}^{a} K^{\{\lambda\},k}_{\{\mu\}\{\nu\}}(\xi_1, \xi_2), & d_{\{\lambda\}} > b. \end{cases}$$

The functions $K^{\{\lambda\},k}_{\{\mu\}\{\nu\}}(\xi_1, \xi_2)$ are given by (4.9), (4.7). The first term in the rhs vanishes if $\xi_1 \neq \xi_2$. The remainder $Q^{(N)}_{\{\mu\}\{\nu\}}(x, \varrho\xi_1, \varrho\xi_2)$ has the order $\varrho^{N-a_1-a_2}$ when $\varrho \to 0$, according to the following lemma.

Lemma 4.2. For every diagram Γ contributing to $Q^{(N)}$ the corresponding FA $G_\Gamma(x, \varrho\xi_1, \varrho\xi_2 \mid \boldsymbol{p})$ regarded at $(\xi_1 - \xi_2)^2 \neq 0$ as a tempered distribution of x and $\boldsymbol{p}$ has $N - a_1 - a_2$ derivatives with respect to ϱ which are continuous for $\varrho \geqq 0$ and vanish at $\varrho = 0$.

A proof of this lemma will be presented in the following Section. It should be noted that expression (4.7) which defines $Q^{(N)}$ possesses the same structure as expression (3.8) for the product of composite fields. Therefore, both $Q^{(N)}$ and $T\big(Sb_1(x + \xi_1)\, b_2(x + \xi_2)\big)$ expand in the same set of diagrams. However, for $Q^{(N)}$, the (sub)diagrams including two vertices $x + \xi_{1,2}$ simultaneously should be subtracted by $\mathfrak{M}_x{}^{a,b}$ rather than by $M^{a,b}$.

2. We conclude this section with the following remarks.

1) A generalization of the light-cone expansion of ref. [13] for theories including massless particles can easily be achieved by replacing the operator $\overline{\mathfrak{M}}^{(a)}$ introduced in [13] by

$$\overline{\mathfrak{M}}^{a,b} = \begin{cases} \overline{\mathfrak{M}}^{(a)} I, & a \leqq b \\ \overline{\mathfrak{M}}^{(b)} I + \sum\limits_{k=b+1}^{a} \Delta^{a-k} (\overline{\mathfrak{M}}^{(k)} - \overline{\mathfrak{M}}^{(k-1)}), & a > b. \end{cases}$$

2) It is worth-while to derive the explicit formulae for the OPE in the MS-scheme [18] which is of great use in practical calculations. In refs. [19, 20] it is described how the corresponding combinatorics may be resolved. However, the analytical part of the proofs is still needed.

5. The proofs

1. In this Section we prove Theorem 2.1 and Lemma 4.2. Both proofs have much in common: they are based on the method developed in ref. [15] and described in the preceding paper. This method takes into account the analogy between UV and IR divergences as far as possible. First, the α-integral is decomposed into sectors and the variables exhibiting the divergences of both kind are introduced. Second, for different terms which contribute to the sector integral under consideration the mixed representation [15] is appropriately chosen. In other words, some of the external vertices are chosen in the momentum space, and the others in the position space. Note that representation (5.24) of the preceding paper plays a decisive role.

The proof of Theorem 2.1 is presented in Subsect. 5.2. It happens to be a part of the proof of Lemma 4.2 because, while analyzing the asymptotic behaviour of FA, one should demonstrate both UV and IR convergence.

2. Let us represent the R-operation as $R = \sum\limits_F \prod\limits_{\gamma \in F} (-M_\gamma{}^{a_\gamma, b_\gamma})$ where the sum runs over the forests of 1 PI subgraphs, and the numbers a_γ and b_γ still fulfill the conditions A) and B). The equivalence of this form and (2.6) will be justified in the end of Subsection 5.2.

Consider now FA $G_\Gamma(\boldsymbol{p}, \mu^2)$ that is regularized by the functions $\varphi(\alpha\mu^2)$ and corresponds to a 1 PI graph Γ

$$G_\Gamma(\boldsymbol{p}, \mu^2) \equiv G_\Gamma{}^\varphi(\boldsymbol{p}) = \int_0^\infty d\alpha G_\Gamma{}^\varphi(\boldsymbol{p}, \alpha)$$

$$= \int_0^\infty d\alpha \prod_{l\in\Gamma} \varphi(\alpha_l) \frac{Z(\tilde{\boldsymbol{p}}, \alpha)}{D^2(\alpha)} \exp\left\{i \frac{A(\tilde{\boldsymbol{p}}, \alpha)}{D(\alpha)} - i \sum_l \tilde{m}_l{}^2\alpha_l\right\}. \tag{5.1}$$

Here $Z(\tilde{\boldsymbol{p}}, \alpha)$ is the function taking into account the derivatives in the Lagrangian and nonscalar character of the propagators:

$$Z(\tilde{\boldsymbol{p}}, \alpha) = \prod_l P_l\left(\frac{1}{2i}\frac{\partial}{\partial\xi_l}\right) \exp\{-iD^{-1}(\alpha)\,[K(\xi, \alpha) + 2B(\tilde{\boldsymbol{p}}, \xi, \alpha)]\}|_{\xi=0}.$$

The definition of the homogeneous forms may be found e.g. in ref. [7]. In particular,

$$D(\alpha) = \sum_{T\in T^{[1]}} \prod_{l\notin T} \alpha_l; \qquad A(\tilde{\boldsymbol{p}}, \alpha) = \sum_{T\in T^{[2]}} \prod_{l\notin T} \alpha_l(\textstyle\sum' \tilde{p}_i)^2.$$

The first sum runs over the trees of the graph while the second sum is extended to 2-trees, $\sum' p_i$ being the total momentum flowing into one of the components of a 2-tree.

The initial variables α_l have the dimension -2 in mass units. In (5.1) it is understood that dimensionless $\alpha_l' = m^2\alpha_l$ are already used and, for simplicity, the parameter m is chosen as IR-regularizing mass μ. Thus, the μ-dependence of functions φ has been disappeared, $\overline{m}_l{}^2 = m_l{}^2/\mu^2$, $\tilde{p} = p/\mu$.*) Let us decompose the integral in rhs of (5.1) into the sum of intcgrals over the regions (sectors)

$$\Delta_{\pi,j} = \{\alpha \mid \alpha_{\pi(1)} \leqq \alpha_{\pi(2)} \leqq \cdots \leqq \alpha_{\pi(j)} \leqq 1 \leqq \alpha_{\pi(j+1)} \leqq \cdots \leqq \alpha_{\pi(L)}\}.$$

Here π is a permutation of indices 1, 2, ..., L. Without loss of generality, we shall consider only the contribution G_Δ corresponding to the sector $\Delta_j = \Delta_{I,j}$ associated with the identical permutation. Furthermore, for $l \geqq j + 1$ we introduce the variables $\beta_l = \alpha_l{}^{-1}$ in Δ_j.

While analyzing different contributions to G_Δ it is useful to make the Fourier transform with respect to some variables $\boldsymbol{p}$, i.e. to apply the mixed representation [15] which, up to a nonessential coefficient has the form

$$\int d\alpha \int d\beta \prod_{l=1}^{j} \varphi(\alpha_l) \prod_{l'=j+1}^{L} \varphi(\beta_{l'}^{-1})\, f^r(\boldsymbol{x}, \boldsymbol{p}, \alpha, \beta) \exp\left\{-i \sum_{l=1}^{j} m_l{}^2\alpha_l - i \sum_{l'=j+1}^{L} m_{l'}^2\beta_{l'}^{-1}\right\} \tag{5.2}$$

where

$$f^r(\boldsymbol{x}, \boldsymbol{p}, \alpha, \beta) = [A^r(\alpha, \beta)]^{-2}\, Z^r(\boldsymbol{x}, \boldsymbol{p}, \alpha, \beta) \exp\left(-iH^r(\boldsymbol{x}, \boldsymbol{p}, \alpha, \beta)\right).$$

As for the definition of A^r, Z^r and H^r, see ref. [15]. In particular, the function Z^r includes all the nonscalar dependence of the diagram.

Let us remind that the parameters $\boldsymbol{\varkappa} = \{\varkappa_\gamma \mid \gamma \in F\}$ and $\boldsymbol{\zeta} = \{\zeta_\gamma \mid \gamma \in F\}$ are introduced in the integrand in order to define the action of subtracting operators. The parameters $\boldsymbol{\varkappa}$ multiply external momenta of a subgraph while $\boldsymbol{\zeta}$ enter to the functions φ. As a result the functions A^r, Z^r and H^r acquire an extra dependence on $\boldsymbol{\varkappa}$ and are,

*) In what follows the tilde will be omitted for brevity.

eventually, expressed in terms of the products

$$\Pi_T(\alpha, \beta, \varkappa) = \prod_{\substack{1 \leqq l \leqq j \\ l \notin T}} \alpha_l \prod_{\substack{j < l' \leqq L \\ l' \in T}} \beta_{l'} \prod_{\gamma \in F} \varkappa_\gamma{}^{2L(\gamma \setminus T)}$$

where the sum runs over trees, 2-trees or pseudo-trees of the graph $\hat{\Gamma}^r$. This graph is obtained from Γ by adjoining the vertex $\hat{v}^r$ that is connected by an auxiliary line with every vertex considered in the position space. In particular, $\hat{\Gamma} = \hat{\Gamma}^n$, n being the total number of external vertices. For the convergence analysis we introduce the "sector" variables $\boldsymbol{t} = (t_1, \ldots, t_j)$ and $\boldsymbol{\tau} = (\tau_{j+1}, \ldots, \tau_L)$ as follows

$$\alpha_l = t_l \ldots t_j, \quad \beta_{l'} = \tau_{j+1} \ldots \tau_{l'}. \tag{5.3}$$

In the sector Δ_j they belong to the segment $[0, 1]$. If a set of indices l and l' is not specified we imply that $l = 1, 2, \ldots, j$; $l' = j + 1, \ldots, L - 1, L$. Note that the sector Δ_j defines the maximal nest $\mathfrak{N}$ composed of the subgraphs $\gamma_l = \{1, 2, \ldots, l\}$ with $l = 1, 2, \ldots, L$.

To establish the convergence of RG_Δ let us rearrange the terms in R according to an equivalence relation. For its definition we first introduce an operation $F \to \bar{F}$ such that $F \subset \bar{F}$. The forest $\bar{F}$ is constructed from F as follows. Let $\gamma \in F$ and let $\upgamma$ be the union of all the elements of F which are strictly contained in γ. Note that $\upgamma$ may be the empty set. Consider the nest of subgraphs $\{\gamma' \mid \gamma' = \upgamma \cup (\gamma_l \cap \gamma), l = 1, 2, \ldots, L, \gamma' \neq \upgamma\}$. It includes $L(\gamma) - L(\upgamma)$ different elements: $\gamma^1 \subset \gamma^2 \subset \cdots \subset \gamma^{L(\gamma) - L(\upgamma)} \equiv \gamma$. Let us introduce an additional notion. A subgraph is UV-irreducible if it consists of a single line or it is 1 PI. Any subgraph is naturally decomposed into UV-irreducible components. Now, for $i = 1, 2, \ldots, L(\gamma) - L(\upgamma) - 1$ we include in F the UV-irreducible component of γ^i that contains the line $\mathcal{L}(\gamma^i \setminus \gamma^{i-1})$. After using all the elements of the forest F we continue the construction procedure analogously with the help of the nest $\{\gamma' \mid \gamma' = F \cup \gamma_l, l = 1, 2, \ldots, j, \gamma' \neq F\}$. Eventually, one has the forest $\bar{F}$.

To define the equivalence relation mentioned above we need complementary notations. Let us denote by Σ the set of self-energy (SE) generalized vertices with the following properties: a) any $\gamma \in \Sigma$ is connected with the rest part of the graph by two massless lines (the subgraph consisting of these lines will be denoted by γ^*); b) the product of unmodified propagators corresponding to γ^* behaves like p^{-4} when $p \to 0$. In fact, such choice of Σ implies SE-insertions into lines associated with scalar or vector massless particles. Furthermore, let $\Sigma_\mathcal{N}$ be the subset of Σ which includes any subgraph that is a connectivity component of some $\gamma_l \in \mathcal{N}$. Finally, we have the equivalence relation.

Definition. The forests F_1 and F_2 are equivalent if

$$\overline{(F_1 \setminus \Sigma_\mathcal{N})} \cup \Sigma_\mathcal{N} = \overline{(F_2 \setminus \Sigma_\mathcal{N})} \cup \Sigma_\mathcal{N}.$$

When analyzing UV and IR convergence of different contributions to the sector integral auxiliary variables will be used. To introduce them we need the notion of the generalized forest. Any UV-forest with the property $\bar{F} = F$ is a maximal UV-forest denoted by $\mathcal{F}_\alpha$. It contains the lines $\mathcal{L}(F) = \{l \mid l \in \gamma \in F\}$. Let us use the lines $\mathcal{L} \setminus \mathcal{L}(F)$ to construct analogously IR-forest $\mathcal{F}_\beta$ ny means of the nest associated with the inverse order of lines $\bar{\mathcal{N}} = \{\gamma_l' \mid \gamma_l' = \{l, l + 1, \ldots, L\}, l = L, L - 1, \ldots, 1\}$ when doing this procedure the UV-irreducibility is replaced by IR-irreducibility: a subgraph γ is IR-(ir)-reducible iff the graph $\hat{\Gamma}/(\Gamma \setminus \gamma)$ is UV-(ir)reducible. Furthermore, the number of loops $\mathcal{N}(\gamma)$ is replaced by the number $\mathfrak{M}(\gamma)$ of cut-sets of the graph $\hat{\Gamma}$ which belong to γ. Note that F_β is obtained from the empty IR-forest. A set $\mathcal{F} = \mathcal{F}_\alpha \cup \mathcal{F}_\beta$ with $\mathcal{L}(F) = \mathcal{L}$ is a maximal generalized forest (or, simply, forest) i.e. a union of UV-forest $\mathcal{F}_\alpha$ and IR-forest $\mathcal{F}_\beta$, the elements of $\mathcal{F}_\alpha$ and $\mathcal{F}_\beta$ having no lines in common. Any $\mathcal{F}$ has exactly L elements. Let us introduce the auxiliary variables $\boldsymbol{t}' = \{t_\gamma' \mid \gamma \in \mathcal{F}_\alpha\}$ and $\boldsymbol{\tau}' = (\tau_\gamma' \mid \gamma \in \mathcal{F}_\beta\}$ associated with $\mathcal{F}$

$$\alpha_l = \prod_{l \in \gamma \in \mathcal{F}_\alpha} t_\gamma', \; l \in \mathcal{L}_\alpha \equiv \mathcal{L}(\mathcal{F}_\alpha); \qquad \beta_l = \prod_{l \in \gamma \in \mathcal{F}_\beta} \tau_\gamma', \; l \in \mathcal{L}_\beta \equiv \mathcal{L}(\mathcal{F}_\beta). \tag{5.4}$$

Then we define the mapping $\sigma\colon \mathcal{F} \to \mathcal{L}$ such that $\sigma(\gamma) \in \mathcal{L}(\gamma)$; $\sigma(\gamma) \notin \mathcal{L}(\gamma')\ \forall\, \gamma' \Subset \gamma$, $\gamma' \in \mathcal{F}$. Furthermore, $\bar{\gamma} \in \mathcal{F}$ is the minimal element of $\mathcal{F}$ with $\bar{\gamma} \Supset \gamma$. The variables $(\boldsymbol{t}', \boldsymbol{\tau}')$ are expressed in terms of $(\boldsymbol{\alpha}, \boldsymbol{\beta})$ as follows: $t_\gamma' = \alpha_{\sigma(\gamma)}$ if γ is a maximal element of $\mathcal{F}_\alpha$ and $t_\gamma' = \alpha_{\sigma(\gamma)}/\alpha_{\sigma(\bar{\gamma})}$ elsewhere; $\tau_\gamma' = \beta_{\sigma(\gamma)}$ if γ is a maximal element of $\mathcal{F}_\beta$ and $\tau_\gamma' = \beta_{\sigma(\gamma)}/\beta_{\sigma(\bar{\gamma})}$ elsewhere. As a result the connection between $(\boldsymbol{t}', \boldsymbol{\tau}')$ and $(\boldsymbol{t}, \boldsymbol{\tau})$ is obtained. When γ is a maximal element (of $\mathcal{F}_\alpha$ or $\mathcal{F}_\beta$) we have

$$\begin{aligned} t_\gamma' &= \begin{cases} t_{\sigma(\gamma)} \dots t_j, & \sigma(\gamma) \leqq j; \\ [\tau_{j+1} \dots \tau_{\sigma(\gamma)}]^{-1}, & \sigma(\gamma) > j; \end{cases} \\ \tau_\gamma &= \tau_{j+1} \dots \tau_{\sigma(\gamma)}, \qquad \sigma(\gamma) > j. \end{aligned} \tag{5.5}$$

If γ is not maximal then

$$\begin{aligned} t_\gamma' &- \begin{cases} t_{\sigma(\gamma)} \dots t_{\sigma(\bar{\gamma})-1}, & \sigma(\gamma) < \sigma(\bar{\gamma}) \leqq j; \\ (t_{\sigma(\bar{\gamma})} \dots t_{\sigma(\gamma)-1})^{-1}, & \sigma(\bar{\gamma}) < \sigma(\gamma) \leqq j; \\ t_{\sigma(\gamma)} \dots t_j \tau_{j+1} \dots \tau_{\sigma(\bar{\gamma})}, & \sigma(\gamma) \leqq j < \sigma(\bar{\gamma}); \\ (t_{\sigma(\bar{\gamma})} \dots t_j \tau_{j+1} \dots \tau_{\sigma(\gamma)})^{-1}, & \sigma(\bar{\gamma}) \leqq j < \sigma(\gamma); \\ \tau_{\sigma(\gamma)+1} \dots \tau_{\sigma(\bar{\gamma})}, & j < \sigma(\gamma) < \sigma(\bar{\gamma}); \\ (\tau_{\sigma(\bar{\gamma})+1} \dots \tau\sigma_{(\gamma)})^{-1}, & j < \sigma(\bar{\gamma}) < \sigma(\gamma); \end{cases} \\ \tau_\gamma' &= \tau_{\sigma(\bar{\gamma})+1} \dots \tau_{\sigma(\gamma)}, \qquad j < \sigma(\bar{\gamma}) < \sigma(\gamma). \end{aligned} \tag{5.6}$$

Let us represent $\mathcal{F}_\alpha$ as the union $\mathcal{F}_\alpha^+ \cup \mathcal{F}_\alpha^-$. An element γ belongs to $\mathcal{F}_\alpha^+$ if either $\sigma(\gamma) < \sigma(\bar{\gamma})$ or it is maximal and $\sigma(\gamma) \leqq j$. It is easy to show that $\mathcal{F}_\alpha = \overline{F \setminus \Sigma_{\mathcal{N}}}$ is constructed from $F \setminus \Sigma_{\mathcal{N}}$ by elements of $\mathcal{F}_\alpha^+$. Thus, any $F \setminus \Sigma_{\mathcal{N}}$ may be obtained from $\mathcal{F}_\alpha$ by removing some of elements of the set $\mathcal{F}_\alpha^+$. Hence, any element of an equivalence class is characterized by $\mathcal{F}_\alpha^-$, a subset of $\mathcal{F}_\alpha^+$ and a subset of $\Sigma_{\mathcal{N}} \setminus \mathcal{F}_\alpha^+$. Therefore, the contribution of the equivalence class to R-operation becomes

$$R_{\mathcal{F},\mathcal{N}} = \prod_{\gamma \in \mathcal{F}_\alpha^-} (-M_\gamma^{a_\gamma, b_\gamma}) \prod_{\gamma \in \mathcal{F}_\alpha^+ \cup \Sigma_{\mathcal{N}}} (1 - M_\gamma^{a_\gamma, b_\gamma}). \tag{5.7}$$

We shall use different representations of operators $M_\gamma^{a_\gamma, b_\gamma}$ via operators $\mathcal{M}$. The definition of $\mathcal{M}$ requires auxiliary parameters $\varkappa_\gamma$ and ζ_γ in G_Δ. Now let us introduce them as follows. The parameters ζ_γ enter to the functions $\varphi\colon \prod_l \varphi(\alpha_l) \Rightarrow \prod_l \varphi\left(\alpha_l\left(1 - \prod_{\gamma \ni l} \zeta_\gamma^2\right)\right)$. As to $\varkappa_\gamma$'s they are introduced as in the case of usual UV subtractions: all the external momenta of a subgraph γ are multiplied by $\varkappa_\gamma$. Eventually, these parameters enter as coefficients of some α_l. Let us now turn to the Fourier transform with respect to some momenta. The choice of these momenta depends on $\mathcal{F}$ and will be described below.

The transformed function $R_{\mathcal{F},\mathcal{N}}G_\Delta(\boldsymbol{p}, \boldsymbol{x})$ becomes

$$R_{\mathcal{F},\mathcal{N}}G_\Delta(\boldsymbol{p}, \boldsymbol{x} \mid \varkappa, \zeta) = \int_0^1 \prod_l t_l^{l-1}\, dt_l \prod_{l'} \tau_{l'}^{L-l'}\, d\tau_{l'}$$
$$\times \exp\Big\{-i\Big[\sum_l m_l^2 \alpha_l(\boldsymbol{t}) + \sum_{l'} m_{l'}^2 \big(\beta_{l'}(\boldsymbol{\tau})\big)^{-1}\Big]\Big\}$$
$$\times \prod_{\gamma \in \mathcal{F} \cup \Sigma_{\mathcal{N}}} \varkappa_\gamma'^{\,4\mathfrak{N}(\gamma)+a(\gamma)} f^r(\boldsymbol{x}, \boldsymbol{p}, \alpha, \boldsymbol{\beta} \mid \varkappa) \prod_l \varphi\Big(\alpha_l(\boldsymbol{t}) \Big(1 - \prod_{\gamma \ni l} \zeta_\gamma^2\Big)\Big)$$
$$\times \prod_{l'} \varphi\Big(\big(\beta_l(\boldsymbol{\tau})\big)^{-1} \Big(1 - \prod_{\gamma \ni l'} \zeta_\gamma^2\Big)\Big). \tag{5.8}$$

Here $a(\gamma) = \sum_{l \in \gamma} a_l$, a_l is the degree of the polynomial P_l in the numerator of l-th propagator which also involves vertex derivatives. The mixed representation [15] is chosen as follows. Include into the set $\mathcal{V}^r$ (associated with position space variables) a vertex from every subset of vertices which are connected with each other by paths composed of lines $\mathcal{L}_\alpha$. As a result the exponent of the corresponding mixed representation becomes an infinitely differentiable function of the variables $(\boldsymbol{t}', \boldsymbol{\tau}')$ [15] and does not involve negative powers of $\varkappa$ due to the definition of $\Sigma_{\mathcal{N}}$.

Now we represent the function f^r from (5.8) as

$$\prod_{l=1}^{j} \alpha_l^{-2} \prod_{l \in \mathcal{L}_\alpha} (\alpha_l')^2 f^r(\boldsymbol{x}, \boldsymbol{p}, \alpha', \boldsymbol{\beta}' \mid \varkappa) = \prod_{l=1}^{j} t_l^{-2l} \prod_{\gamma \in \mathcal{F}_\alpha} (t_\gamma')^{2L(\gamma)} f^r(\ldots, t', \boldsymbol{\tau}', \ldots). \tag{5.9}$$

The function $f^r(\ldots, \alpha', \beta', \ldots)$ is expressed in terms of the polylinear forms associated with the decomposition of the set $\mathcal{L}$ into $\mathcal{L}_\alpha$ and $\mathcal{L}_\beta = \mathcal{L} \setminus \mathcal{L}_\alpha$. Apply then the factorization of this function in the variables $(\boldsymbol{t}', \boldsymbol{\tau}')$ corresponding to the maximal forest $\mathcal{F}$ (see formula (5.24) of the preceding paper). The factorization is easily generalized when the parameters $\varkappa$ are involved

$$f^r(\boldsymbol{x}, \boldsymbol{p}, \boldsymbol{t}', \boldsymbol{\tau}' \mid \varkappa) = \prod_{\gamma \in \mathcal{F}_\alpha} (\varkappa_\gamma^2 t_\gamma')^{-[\omega(\gamma)/2] - L(\gamma)}$$
$$\times \prod_{\gamma \in \mathcal{F}_\beta} (\tau_\gamma')^{-\Omega(\gamma) - L(\gamma)} \prod_{\gamma \in \Sigma_{\mathcal{N}} \setminus \mathcal{F}} \varkappa_\gamma^{-2[\omega(\gamma)/2] - 2L(\gamma)} Q(\boldsymbol{x}, \boldsymbol{p}, \boldsymbol{t}' \mid \boldsymbol{\tau}' \mid \varkappa). \tag{5.10}$$

Here $Q \in C^\infty$, and $[x]$ always stands for the integer part of a number. Furthermore, Ω is IR index of divergence that satisfies the inequality

$$\Omega(\gamma) \leqq \Big[1/2\big(\omega(\widehat{\Gamma \setminus \gamma}) - \omega(\hat{\Gamma})\big) + \sum_i \operatorname{Odd}\, (\Gamma \setminus \gamma)_i + 1)\Big]$$

where $(\Gamma \setminus \gamma)_i$ are the connectivity components of the subgraph $\Gamma \setminus \gamma$, and the oddness of γ equals $a(\gamma) - 2[a(\gamma)/2]$.

The variable t_γ' occurs in Q only in the factor $\varkappa_\gamma^2 t_\gamma' \pi_\gamma(\boldsymbol{\tau})$ where $\pi_\gamma(\boldsymbol{\tau}) = \prod \tau_{l'}$, and the products goes over l' with $\gamma \subset \gamma_{l'-1}$, $\gamma^* \cap \gamma_{l'-1} = \emptyset$. As for $\varkappa_\gamma$ at $\gamma \in \Sigma_{\mathcal{N}} \setminus \mathcal{F}$, it occurs in Q only in the factors $\varkappa_\gamma^2 \pi_\gamma(\boldsymbol{\tau})$. These statements follow from the factorization properties of the forms and from the fact that $\operatorname{def}_T \gamma \leqq \operatorname{def}_T \gamma_{l'-1}$ for any γ with $\gamma \subset \gamma_{l'-1}$ and $\gamma^* \cap \gamma_{l'-1} = \emptyset$ (here $\operatorname{def}_T \gamma = c(\gamma \cap T) - c(\gamma)$ is the "defect" [7], $c(\gamma)$ is the number of the connectivity components of γ). Note also that the parameters ζ_γ at $\gamma \in \mathcal{F}_\alpha$ occur in the functions φ in the factors $t_\gamma' \zeta_\gamma^2$ only.

Relations (5.5), (5.6) show that t_γ' involves negative powers of the variables $(\boldsymbol{t}, \boldsymbol{\tau})$ if, and only if $\gamma \in \mathcal{F}_\alpha^-$. However, after the action of the operator $M_\gamma^{a_\gamma, b_\gamma}$ the corresponding parameter $\varkappa_\gamma$ vanishes.

Therefore, representation (5.10) provides a factorization in $(\boldsymbol{t}, \boldsymbol{\tau})$ as well. In other words, rhs of (5.10) takes the form $\prod t_l^{N_l} \prod \tau_{l'}^{\bar{N}_{l'}} \Psi(\boldsymbol{t}, \boldsymbol{\tau})$ with $\Psi \in C^\infty$ due to operating of $M_\gamma^{a_\gamma, b_\gamma}$ for $\gamma \in \mathcal{F}_\alpha^-$.

Let us prove the UV convergence of the integral $R_{\mathcal{F},\mathcal{N}} G_\Delta$. First, we turn in (5.7) to the operators $\mathcal{M}$ which act on parameters. We define the parameters $(\varkappa, \zeta)$ introduced by replacing $(\varkappa, \zeta)$ by $(\varkappa'\zeta', \zeta')$. Then, with primes omited, we have the correspondence

$$M_\gamma^{a_\gamma, b_\gamma} \Rightarrow \mathcal{M}_{\varkappa_\gamma}^{b_\gamma} I_{\zeta_\gamma} + (I_{\varkappa_\gamma} - M_{\varkappa_\gamma}^{b_\gamma}) \mathcal{M}_{\zeta_\gamma}^{a_\gamma}. \tag{5.11}$$

Using the standard representation for $I_\zeta - M_\zeta^a$

$$(I_\zeta - \mathcal{M}_\zeta^a) f(\zeta) = \int_0^1 d\zeta \, \frac{(1-\zeta)^a}{a!} \left(\frac{\partial}{\partial \zeta}\right)^{a+1} f(\zeta) \tag{5.12}$$

we evaluate the powers of t_l in the integrand of $R_{\mathcal{F},\mathcal{N}} G_\Lambda$. In accordance with (5.5)–(5.10) we have

$$\deg t_l \equiv N_l = -l - 1 + \sum_{\sigma(\gamma) \leqq l < \sigma(\bar{\gamma})} N_\gamma^+ - \sum_{\sigma(\bar{\gamma}) \leqq l < \sigma(\gamma)} N_\gamma^- \tag{5.13}$$

where $N_\gamma^\pm$ are the powers of t_γ' for $\gamma \in \mathcal{F}_\alpha^\pm$. Furthermore, the condition $\sigma(\gamma) \leqq l < \sigma(\bar{\gamma})$ is replaced by $\sigma(\gamma) \leqq l$ for maximal elements γ. Formulae (5.8), (5.10) show that the parameter ζ_γ has the power $\omega(\gamma) - 2[\omega(\gamma)/2]$ which reduces the degree of the operator $\mathcal{M}_\zeta^a$ because $\mathcal{M}_{\zeta_\gamma}^{a_\gamma} \zeta_\gamma^{\omega(\gamma) - 2[1/2\omega(\gamma)]} = \mathcal{M}_{\zeta_\gamma}^{a^\gamma + 2[1/2\omega(\gamma)]}$. Due to specific dependence of the function Q, the action of the derivative $(\partial/\partial\zeta_\gamma)^k$ leads to the auxiliary power $[1/2(k+1)]$ of the variable t_γ'. But if the parameter ζ_γ is nullified $(\gamma \in \mathcal{F}_\alpha^-)$ then the auxiliary power becomes $[1/2k]$. Taking into account the powers of t_γ' in (5.9) and (5.10) we have

$$N_\gamma^- \leqq [a^\gamma/2] + L_\gamma. \tag{5.14}$$

Similarly, in accordance with (5.11), the action of $1 - M_\gamma^{a_\gamma, b_\gamma}$ reduces to the differentiation of the order $a_\gamma + 1 \equiv \omega_\gamma + a^\gamma + 1$ in ζ_γ when power counting is considered. Thus, for $\gamma \in \mathcal{F}_\alpha^+$,

$$N_\gamma^+ \geqq [a^\gamma/2] + L_\gamma + 1. \tag{5.15}$$

Rearranging the terms in (5.13) with the help of the relations

$$\sum_{\sigma(\gamma) \leqq l < \sigma(\bar{\gamma})} = \sum_{\sigma(\gamma) \leqq l} - \sum_{\sigma(\gamma), \sigma(\bar{\gamma}) \leqq l}, \qquad \sum_{\sigma(\bar{\gamma}) \leqq l < \sigma(\gamma)} = \sum_{\gamma : \sigma(\bar{\gamma}) \leqq l} - \sum_{\sigma(\gamma), \sigma(\bar{\gamma}) \leqq l},$$
$$\sum_{\gamma : \sigma(\bar{\gamma}) \leqq l} A_\gamma = \sum_{\gamma : \sigma(\gamma) \leqq l} \sum_{\gamma' : \bar{\gamma}' = \gamma} A_\gamma \tag{5.16}$$

applying (5.13) and taking (5.14) and (5.15) into account we obtain the inequality $N_l \geqq \sum_{\substack{\sigma(\gamma) \leqq l < \sigma(\bar{\gamma}) \\ \gamma \in \mathcal{F}_\alpha}} 1 - 1$, provided degrees a^γ satisfy the conditions of the theorem. Let us show that $N_l \geqq 0$ for any $1 \leqq l \leqq j$, i.e. there always exists $\gamma \in \mathcal{F}_\alpha$ with $\sigma(\gamma) \leqq l < \sigma(\bar{\gamma})$. Suppose that it is not true. Then $\sigma(\gamma) \geqq l + 1$ for all maximal elements of $\mathcal{F}_\alpha$. Let $\{\gamma^i\}$ be pairwise disjoint elements of $\mathcal{F}_\alpha$ which are in some maximal element. According to the assumption, $\sigma(\gamma^i) \geqq l + 1$. By continuing the consideration analogously we get that $\sigma(\gamma) \geqq l + 1$ for all $\gamma \in \mathcal{F}_\alpha$. Since $\sigma(\gamma) \geqq j + 1$ for $\gamma \in \mathcal{F}_\beta$, due to the construction procedure of $\mathcal{F}_\beta$, we have $\sigma(\gamma) \geqq l + 1 \; \forall \gamma \in \mathcal{F}$. However, as $\mathcal{F}$ contains exactly L elements, this contradicts the assumption. Thus, $N_l \geqq 0$.

Let us turn to IR convergence. Now we introduce parameters $\varkappa_\gamma$ and ζ_γ by the initial way and use the following representation for $M_\gamma{}^{a_\gamma, b_\gamma}$

$$M_\gamma{}^{a_\gamma,b_\gamma} \Rightarrow \mathcal{M}^{b_\gamma}_{\varkappa_\gamma} I_{\zeta_\gamma} + \sum_{k_\gamma=b_\gamma+1}^{a_\gamma} \mathcal{M}^{a_\gamma-k_\gamma}_{\zeta_\gamma}(\mathcal{M}^{k_\gamma}_{\varkappa_\gamma} - \mathcal{M}^{k_\gamma-1}_{\varkappa_\gamma}).$$

Note that the integrand in $R_{\mathcal{F},\mathcal{N}} G_\Delta$ is a sum of terms which (up to infinitely differentiable function) look like products consisting of the variables τ and t in some powers, the functions φ and their derivatives, and the massive exponent. Therefore, the convergence with respect to any $\tau_{l'}$ relies on non-negativity of its power if the term considered does not involve the functions φ and/or the massive exponent depending on the given $\tau_{l'}$. In other cases the IR convergence is provided by these functions: this fact will be shown below.

Let $\gamma \in \mathcal{F}_\alpha^-$. If the action of the operator $\mathcal{M}^{(k_\gamma)}_{\varkappa_\gamma}$ changes the power of $\tau_{l'}$ then there is a line in γ with the number that is not less than l' (see relations (5.5), (5.6)). However, in this case, the action of the operators $\mathcal{M}^{a_\gamma-k_\gamma}_{\zeta_\gamma}$ leads to appearence of functions φ or their derivatives depending on $\tau_{l'}$. Therefore, it is sufficient to consider only the contribution $\mathcal{M}^{b_\gamma}_{\varkappa_\gamma} I_{\zeta_\gamma}$ into $M_\gamma{}^{a_\gamma,b_\gamma}$ in accordance with the previous remark.

Let now $\gamma \in \mathcal{F}_\alpha^+ \cup \Sigma_\mathcal{N}$. Represent $1 - M_\gamma{}^{a_\gamma,b_\gamma}$ as $(I_{\varkappa_\gamma} - \mathcal{M}^{b_\gamma}_{\varkappa_\gamma}) I_{\zeta_\gamma} + \sum_{k_\gamma=b_\gamma+1}^{a_\gamma} \mathcal{M}^{a_\gamma-k_\gamma}_{\zeta_\gamma} \mathcal{M}^{(k_\gamma)}_{\varkappa_\gamma}$.
If the operators $\mathcal{M}^{a_\gamma-k_\gamma}_{\zeta_\gamma}$ provide the convergence with respect to $\tau_{l'}$ then it suffices to consider only the contribution $(I_{\varkappa_\gamma} - \mathcal{M}^{b_\gamma}_{\varkappa_\gamma}) I_{\zeta_\gamma}$ in $1 - M_\gamma{}^{a_\gamma,b_\gamma}$. Otherwise, it is necessary to take into account the rest part as well. Note that any term in this sum leads to the differentiation $(\partial/\partial\varkappa_\gamma)^{k_\gamma}$ of the order $k_\gamma \geqq b_\gamma + 1$. On the other hand, formula (5.12) for the remainder shows that the action of $I_{\varkappa_\gamma} - \mathcal{M}^{b_\gamma}_{\varkappa_\gamma}$ reduces to application of the latter operation when power counting of $\tau_{l'}$ is considered. Eventually, we have

$$\deg \tau_{l'} = Q_{l'} + L - l' + \sum_{\substack{\sigma(\bar\gamma)<l'\leqq\sigma(\gamma),\\ \gamma\in\mathcal{F}_\beta}} \deg \tau_\gamma' + \sum_{\substack{\sigma(\gamma)<l'\leqq\sigma(\bar\gamma),\\ \gamma\in\mathcal{F}_\alpha}} N_\gamma'^{+} - \sum_{\substack{\sigma(\bar\gamma)<l'\leqq\sigma(\gamma),\\ \gamma\in\mathcal{F}_\alpha}} N_\gamma'^{-}. \tag{5.17}$$

Here $Q_{l'}$ is the contribution to the power of $\tau_{l'}$ from the action of $\prod_{\gamma\in\Sigma_\mathcal{N}} (I_{\varkappa_\gamma} - \mathcal{M}^{b_\gamma}_{\varkappa_\gamma})$ due to the factor $\pi_\gamma(\tau)$. This contribution looks like

$$Q_{l'} = \sum_{\substack{\gamma\in\Sigma_\mathcal{N},\, \gamma\subset\gamma_{l'-1},\\ \gamma^*\cap\gamma_{l'-1}=\emptyset}} ([1/2b_\gamma] - \text{Odd}\,\gamma + 1). \tag{5.18}$$

Formulae (5.10) and (5.11) yield

$$\deg \tau_{l'} \geqq L - l' + \sum_{\substack{\sigma(\gamma)<l'\leqq\sigma(\bar\gamma),\\ \gamma\in\mathcal{F}_\alpha}} \big(L(\gamma) - [1/2b^\gamma]\big) - \sum_{\substack{\sigma(\bar\gamma)<l'\leqq\sigma(\gamma),\\ \gamma\in\mathcal{F}_\alpha}} \big(L(\gamma) - [1/2\,(b^\gamma + 1)]\big)$$
$$- \sum_{\substack{\sigma(\bar\gamma)<l'\leqq\sigma(\gamma),\\ \gamma\in\mathcal{F}_\beta}} \big(\Omega(\gamma) + L(\gamma)\big) + Q_{l'}$$

with $b^\gamma = \omega_\gamma - 1 - b_\gamma$. With the help of the relations which are analogous to (5.16) one comes to

$$\deg \tau_{l'} \geqq L - l' + \sum_{\substack{\sigma(\bar\gamma)<l'\leqq\sigma(\bar\gamma),\\ \gamma\in\mathcal{F}_\beta}} 1 - 1 + \sum_{\substack{\sigma(\gamma)\geqq l,\\ \gamma\in\mathcal{F}_\alpha}} \Big([1/2(b^\gamma + 1)] - \sum_{\gamma':\bar\gamma'=\gamma} [1/2b^{\gamma'}]\Big)$$
$$- \sum_{\substack{\sigma(\bar\gamma)<l'\leqq\sigma(\gamma),\\ \gamma\in\mathcal{F}_\beta}} \big(\Omega(\gamma) + 1\big) + Q_{l'}. \tag{5.19}$$

The sum of the two initial terms in rhs is non-negative: this is verified by the same arguments as in the case of deg t_l. The third term is also non-negative due to conditions B) of the theorem. To consider the last terms let us use the nature of IR divergences in theories involved. The analysis of ref. [7] shows that $\Omega(\gamma) \geqq 0$ only if the graph $\Gamma \setminus \gamma$ includes some $\gamma' \in \Sigma_{\mathcal{N}}$ as a connectivity component. However, in this case the cancellation occurs between the last sums in (5.19). Thus, deg $\tau_{l'} \geqq 0$ for l' such that $\Gamma \setminus \gamma_{l'-1}$ has not any line which implies regularization by means of the functions φ.

To complete the proof of the IR convergence it is necessary to show that the dependence of the function φ on $\tau_{l'}$ provides the convergence in $\tau_{l'}$ irrespective of deg $\tau_{l'}$. Note, first, that the convergence is obviously provided by rapidly decreasing functions. In fact, some "oscillated" functions, e.g. massive exponent $\exp(-i\mu^2\alpha)$, are also applicable. To establish the IR convergence it suffices to show that the limit exists when some auxiliary IR regularization (for instance, dimensional or analytic one) is removed. To do this, one may use either the standard technique of ref. [7] or the intergation by parts in the variables $\tau_{l'}$ [15]. In the first case the function φ should rapidly decrease in below halfplane of the complex argument. In the second case it suffices to suppose (a closely related condition) that

$$\left|\left(\frac{\chi^{(n)}}{\chi^{(n+1)}}\right)^{(m)}\right| < C_{m,n}\beta^{\eta-m+1}, \qquad \beta \leqq \beta_0 \tag{5.20}$$

for some η, $C_{m,n}$, $\beta_0 > 0$ and every $m, n \geqq 0$. Here

$$\chi^{(n)}(\beta) = \frac{d^n}{d\beta^n}\,\varphi(\beta^{-1}).$$

Let us briefly describe the second case. The sector integral is represented as a sum of integrals $\int\limits_0^1 d\boldsymbol{\tau}\chi^{(n_j+1)}(\tau_{j+1}) \dots \chi^{(n_L)}(\tau_{j+1} \dots \tau_L)\,\Psi(\boldsymbol{\tau})$ with $\Psi \in C^\infty$. Here non-essential integration in $\boldsymbol{t}$ is omitted. For simplicity we suppose that $m_l = 0$ for any l, otherwise it would be necessary to compare regularizing properties of the function φ and $\exp(-im_l^2 \times \alpha_l)$. First, we integrate several times by parts over τ_L by choosing the factor $(\chi^{(n_L)})'\,d\tau_L$ in the integrand and applying (5.20). Correspondingly, the powers of all the variables $\tau_{j+1}, \dots, \tau_L$ are each time increased by η. Thus, one can obtain the powers sufficiently large to provide the absolute convergence. It is necessary to consider also the boundary terms in the integration by parts formula. The first of them looks like $\int\limits_0^1 d\tau_{j+1} \dots d\tau_{L-1} \times \chi^{(n_j+1)}(\tau_{j+1}) \dots \chi^{(n_{L-1})}(\tau_{L-1})\,\chi^{(n_L)}(\tau_{L-1})$. We use (5.20) again. Suppose, for definitness, that $n_L \leqq n_{L-1}$. Then we choose the factor $(\chi^{(n_L-1)})^2$ in the integrand. Note that inequalities (5.20) hold also for the powers $(\chi^{(n)})^m$ of derivatives $\chi^{(n)}$. With the help of the integration by parts over τ_{L-1} (each time the factor $[(\chi^{(n_{L-1})})^2]'\,d\tau_{L-1}$ is chosen) the powers of $\tau_{j+1}, \dots, \tau_{L-1}$ become sufficiently large. As to boundary terms they lead to the integrals of the type $\int\limits_0^1 d\tau_{j+1} \dots d\tau_{L-2} \dots \chi^{(n_{L-2})}(\tau_{L-2})\left(\chi^{(n_{L-1})}(\tau_{L-2})\right)^2 \dots$ which are analogously studied by means of (5.20).

To complete the proof of the theorem it is necessary to establish the equality

$$R' = \sum_{F_0 \in \{F_0\}} \prod_{\gamma \in F_0} (-M_\gamma^{a_\gamma,b_\gamma}) = \sum_{F \in \{F\}} \prod_{\gamma \in F} (-M_\gamma^{a_\gamma,b_\gamma})$$

where F_0 and F stand for the forests of generalized vertices and 1 PI subgraphs, respectively. Consider a subgraph $\gamma \in F$ that contributes to the second sum and is not complete. Then there exists the subgraph $\gamma' \in \{F_0\}$ which is obtained from γ by adjoining any line $l \notin \gamma$ that connects the vertices of γ. Apply then representation (5.11). One easily verifies that the corresponding operators occur in $\sum_{\{F\}}$ in the combinations $M_\gamma^{a_\gamma, b_\gamma} \times \left(1 - M_{\gamma'}^{a_{\gamma'}, b_{\gamma'}}\right)$. In accordance with (2.2) we have

$$\left(1 - M_{\gamma'}^{a_{\gamma'}, b_{\gamma'}}\right) = \left(1 - M_{\gamma'}^{b_{\gamma'}}\right) \left(I_\gamma + \sum_{k+b_{\gamma'}+1}^{a_{\gamma'}} \Delta_{\gamma'}^{a_{\gamma'}-k} (M_\gamma{}^k - M_\gamma{}^{k-1}\right).$$

Now we choose a_γ and b_γ to sutisfy the inequalities $b_{\gamma'} \geqq a_\gamma \geqq b_\gamma$. Such choice is possible if $\omega_{\gamma'} - b_{\gamma'} - 1$ is not too large. In particular, $b_{\gamma'} = \omega_{\gamma'} - 1$ is always permissible. By analyzing the functional dependence on the parameters $(\varkappa_\gamma, \zeta_\gamma)$ and $(\varkappa_{\gamma'}, \zeta_{\gamma'})$ of the given FA one shows that $(1 - M_{\gamma'}^{b_{\gamma'}}) M_\gamma{}^{b_\gamma} = 0$ and

$$(1 - M_{\gamma'}^{b_{\gamma'}}) M_\gamma{}^k = 0, \qquad b_\gamma < k \leqq a_\gamma.$$

Thus, the terms associated with non-complete subgraphs vanish.

3. Let us prove Lemma 4.2. Without loss of generality we set $x = 0$. Let $G_\Gamma{}^R(\xi_1, \xi_2, \boldsymbol{p})$ be FA of some diagram Γ that contributes to the functional $Q^{(N)} \equiv Q^{(N)}_{\{\mu\}\{\nu\}}(\xi_1, \xi_2)$. We represent $G_\Gamma{}^R$ as $R_\xi G_\Gamma$ with

$$R_\xi = \sum_F \prod_{\gamma \in F} (-\mathfrak{M}_\gamma{}^{a_\gamma, b_\gamma})$$

and use the α-representation

$$G_\Gamma{}^R(\xi_1, \xi_2, \boldsymbol{p}) = R_\xi \int d\alpha G_\Gamma(\xi_1, \xi_2, \boldsymbol{p} \mid \alpha) \exp\left[-i \sum_l m_l{}^2 \alpha_l\right]. \tag{5.21}$$

The sum goes over subgraphs which become 1 PI after contracting the vertices ξ_1, ξ_2. These subgraphs do not necessary involve $\xi_{1,2}$ and may be 1 PI before contracting as well. Let γ^0 be the generalized vertex corresponding to the vertices $\xi_{1,2}$ if there are lines connecting them, and let γ^0 be the subgraph consisting of these isolated vertices, otherwise. By definition, γ^0 contributes into R_ξ and does not overlap with any other subgraph. The operator $\mathfrak{M}_\gamma{}^{a_\gamma, b_\gamma}$ in R_ξ is defined like $\mathfrak{M}^{a,b}$, but, on a (sub)diagram γ rather than on a functional. This is a manifestation of the fact that the functional $Q^{(N)}$ expands in the same set of diagrams as $E_0(s_r)\, 'B^{a_1,b_1}_{\{\mu\}}(\xi_1)\, 'B^{a_2,b_2}_{\{\nu\}}(\xi_2)$; however, for subgraphs which do not include $\xi_{1,2}$ simultaneously, $\mathfrak{M}_\gamma{}^{a_\gamma, b_\gamma}$ should be replaced by the usual operator $M_\gamma{}^{a_\gamma, b_\gamma}$. The function $G_\Gamma(\xi_1, \xi_2, \boldsymbol{p} \mid \alpha)$ in (5.21) looks like [13]

$$\begin{aligned} G_\Gamma(\xi_1, \xi_2, \boldsymbol{p} \mid \alpha) = Z(\xi_1, \xi_2, \boldsymbol{p} \mid \alpha) \exp\Bigg\{&-i \frac{D(\alpha)}{A(\alpha)} \frac{(\xi_1 - \xi_2)^2}{4} \\ &- i \frac{\xi_1 - \xi_2}{A(\alpha)} \sum_i B_i(\alpha)\, p_i - i \frac{\left(\sum_i B_i(\alpha)\, p_i\right)^2}{A(\alpha)\, D(\alpha)} \\ &+ i \frac{C(\alpha, \boldsymbol{p})}{D(\alpha)} + i\xi_2 \sum_i p_i\Bigg\}. \end{aligned}$$

The definitions of the polylinear forms in the α-representation see in ref. [13]. According to the notations of the mixed representation [15] the exponent becomes

$$\begin{aligned} \exp(-iH^2) = \exp\Bigg\{-i \Big[&1/4 U_{12}^2 (\xi_1 - \xi_2)^2 \\ &- \sum_i A_i{}^{2,1} p_i (\xi_1 - \xi_2) - \sum_{i,i'} A_{ii'}^2 p_i p_{i'} - \xi_2 \sum_i p_i\Big] (A^2)^{-1}\Bigg\}. \end{aligned}$$

In particular, $D = U_{12}^2$, $A = A^2$, $B_i = A_i{}^{2,1}$.

Note that in R_ξ the factor $1 - \mathfrak{M}_0$ with $\mathfrak{M}_0 \equiv \mathfrak{M}_{\gamma^0}^{N-a_1-a_2,b-b_1-b_2}$ is picked out. The operator $\mathfrak{M}_0$ acts on the parameter $\varkappa_0$ that occurs only as a coefficient at ξ_1, ξ_2. Thus, the factor involved provides the zero of the order $N - a_1 - a_2$ if FA $G_{I'}{}^R(\xi_1, \xi_2, \boldsymbol{p})$, with the contribution of γ^0 into R_ξ omited, is sufficiently smooth. It is this smoothness that will be established below.

To analyze the behaviour of $G_\Gamma{}^R$ in a vicinity of the point $\xi_1, \xi_2 = 0$ and to prove the convergence of the α-representation we first repeat the initial steps of Subsection 5.2. Integral (5.21) is decomposed into the sum of integrals over the sectors $\Delta_{\pi,j}$, the contribution $R_\xi G_\Delta$ from the sector $\Delta_j \equiv \Delta_{I,j}$ being only considered. Then one turns to sector variables (5.3).

For the convergence analysis of $R_\xi G_\Delta$ it is useful to rearrange the terms in R_ξ according to the equivalence relation similar to that of Subsection 5.2. The difference concerns the change $\Gamma \to \check{\Gamma}$ where $\check{\Gamma}$ is the graph obtained from Γ by contracting the vertices ξ_1 and ξ_2. Two forests $F_{1,2}$ are regarded as equivalent if

$$(\overline{F_1 \setminus \Sigma_{\mathcal{N}}}) \cup \Sigma_{\mathcal{N}} = (\overline{F_2 \setminus \Sigma_{\mathcal{N}}}) \cup \Sigma_{\mathcal{N}}$$

(the definitions of the operation $F \to \bar{F}$ and the sets Σ, $\Sigma_{\mathcal{N}}$ are given in Subsection 5.2). As a result R_ξ transforms into the sum over the equivalence classes characterized by the sets $\mathcal{F}_\alpha{}^-$. The contribution of an equivalence class to R_ξ becomes

$$R_{\mathcal{F},\mathcal{N}} = \prod_{\gamma \in \mathcal{F}_\alpha{}^-} (-\mathfrak{M}_\gamma{}^{a_\gamma,b_\gamma}) \prod_{\gamma \in \mathcal{F}_\alpha{}^+ \cup \Sigma_{\mathcal{N}}} (1 - \mathfrak{M}_\gamma{}^{a_\gamma,b_\gamma}).$$

The action of the subtracting operators is defined, as usual, with the help of the auxiliary parameters $\varkappa$ and ζ. Furthermore, the variables $\xi_{1,2}$ are multiplied by $\prod_{\gamma \in \Gamma_c} \varkappa_\gamma$. Here Γ_c is the set of subgraphs containing both vertices $\xi_{1,2}$ in the same connectivity component.

Now we turn to the Fourier transform of $R_{\mathcal{F},\mathcal{N}} G_\Delta$ with respect to some momenta. The choice of these will be further described. The transformed function $R_{\mathcal{F},\mathcal{N}} G_\Delta(\varrho\xi_1, \varrho\xi_2, \boldsymbol{x}, \boldsymbol{p} \mid \varkappa, \zeta)$ has the form similar to (5.8) ($\xi_{1,2}$ is also multiplied by ϱ)

$$\begin{aligned} &R_{\mathcal{F},\mathcal{N}} \int_0^1 \prod_l t_l{}^{l-1}\, dt_l \prod_{l'} \tau_{l'}^{L-l'}\, d\tau_{l'} \exp\left\{-i \sum_l m_l{}^2 \alpha_l(\boldsymbol{t}) - i \sum_{l'} m_{l'}^2 (\beta_{l'}(\boldsymbol{\tau}))^{-1}\right\} \\ &\times \prod_{\gamma \in \mathcal{F} \cup \Sigma_{\mathcal{N}}} \varkappa_\gamma{}^{4\check{\mathfrak{N}}(\gamma) \pm a(\gamma)} f^r(\varrho\xi_1, \varrho\xi_2, \boldsymbol{x}, \boldsymbol{p}, \alpha, \beta \mid \varkappa) \\ &\times \prod_l \varphi\left(\alpha_l(\boldsymbol{t}) \left(1 - \prod_{\gamma \ni l} \zeta_\gamma{}^2\right)\right) \prod_{l'} \varphi\left((\beta_{l'}(\boldsymbol{\tau}))^{-1} \left(1 - \prod_{\gamma \ni l'} \zeta_\gamma{}^2\right)\right) \end{aligned} \tag{5.22}$$

with $f^r = (A^r)^{-2} Z^r \exp(-iH^r)$ as before.

To analyze the convergence and the asymptotic behaviour at $\varrho \to 0$ of the integral $R_{\mathcal{F},\mathcal{N}} G_\Delta$ let us choose for given $\mathcal{L}_\alpha \equiv \mathcal{L}(\mathcal{F}_\alpha)$ the mixed representation (i.e. the set of momenta mentioned) as follows. Include into $\mathcal{V}^r$ (associated with position space variables) a vertex from every subset of vertices which are conneccted with each other by paths composed of lines $\mathcal{L}_\alpha$. The only exception concerns the veritices $\xi_{1,2}$ which will be always considered in the position space. As a result, provided such choice is done, the exponent H^r of representation (5.22) will not be continuous if these vertices are connected in Γ by a path composed of lines $\mathcal{L}_\alpha$. Specifically, the only term $U_{12}^r (\varrho\xi)^2/A^r$ in H^r will be discontinuous (here A^r and U_{12}^r are the sums of the products Π_T: the former runs over the trees of $\hat{\Gamma}^r$, and the latter runs over the pseudo-trees of $\hat{\Gamma}^r$ including a circuit going through $\xi_{1,2}$ [15]).

We represent the function f^r in (5.22) as

$$\prod_{l=1}^{j} \alpha_l^{-2} \prod_{l \in \mathcal{L}_\alpha} (\alpha_l')^2 f^r(\ldots, \alpha', \beta', \ldots).$$

The function $f^r(\ldots, \alpha', \beta', \ldots)$ is expressed via polylinear forms corresponding to the decomposition of the set $\mathcal{L}$ into $\mathcal{L}_\alpha$ and $\mathcal{L}_\beta$. Let us now use the factorization of this function with respect to the auxiliary variables $(\boldsymbol{t}', \boldsymbol{\tau}')$ associated with a maximal forest $\mathcal{F}$ and related to the initial sector variables by means of (5.4)—(5.6). This factorization has almost the same form as (5.10). The difference concerns the negative powers of the variables $\{t_\gamma' \mid \gamma \in \mathcal{F}_\alpha \cap \Gamma_c\}$ in the exponent, and the changes $\mathcal{N}(\gamma) \to \mathcal{N}(\check{\gamma})$ (the subgraphs $\check{\gamma}$ are obtained like $\check{\Gamma}$) and $\xi_{1,2} \to \varrho\xi_{1,2}$. One easily verifies (by means of representation (14)—(17) of ref. [15]) that these powers occur only in the term $(U_{12}^r/A^r)\,\varrho^2\xi^2$. By the standard procedure based on an appropriate choice of the tree that provides the minimum of the defects, we derive the following representation

$$\frac{U_{12}^r}{A^r} = \frac{1 + P_1(\boldsymbol{t}', \boldsymbol{\tau}')}{1 + P_2(\boldsymbol{t}', \boldsymbol{\tau}')} \cdot \frac{1}{t_{\gamma^1}' \ldots t_{\gamma^k}'}, \tag{5.23}$$

where $P_{1,2}$ are polynomials with non-negative coefficients, and $\gamma^1 \subset \gamma^2 \subset \cdots \subset \gamma^k \in \mathcal{F}_\alpha \cap \Gamma^c \equiv \mathcal{F}_\alpha{}^c$. The set $\mathcal{F}_\alpha{}^c$ is nothing but a nest, therefore, $t_{\gamma^1}' \ldots t_{\gamma^k}' = t_{l_1} \ldots t_j$ with $l_1 = \sigma(\gamma^1)$ (the definition of the mapping σ was given after formula (5.4)). It should be emphasized that representation (5.23) holds due to the special choice of mixed representation (5.22).

Eventually, the factorization of the function $f(\ldots, \alpha, \beta, \ldots)$ takes the form

$$\begin{aligned} f^r(\ldots, \alpha, \beta, \ldots) &= \prod_{l=1}^{j} t_l^{2l} \prod_{\gamma \in \mathcal{F}_\alpha} (t_\gamma')^{2L(\gamma)} f^r(\ldots, \alpha', \beta', \ldots) \\ f^r(\ldots, \alpha', \beta', \ldots) &= \prod_{\gamma \in \mathcal{F}_\alpha} (\varkappa_\gamma{}^2 t_\gamma')^{-[1/2\omega(\gamma)]-L(\gamma)} \prod_{\gamma \in \mathcal{F}_\beta} (\tau_\gamma')^{-\Omega(\gamma)-L(\gamma)} \\ &\quad \times \prod_{\gamma \in \Sigma_{\mathcal{N}} \setminus \mathcal{F}} \varkappa_\gamma{}^{2[1/2\omega(\gamma)]-2L(\gamma)} u(\varrho\xi_1, \varrho\xi_2, \boldsymbol{x}, \boldsymbol{p}, \boldsymbol{t}', \boldsymbol{\tau}' \mid \varkappa), \\ u &= u_1 u_2 u_3; \qquad u_1 = \exp\left(-i\varrho^2\xi^2 V_1 (t_{l_1} \ldots t_j)^{-1}\right), \\ u_2 &= \exp\left(-i\varrho\xi V_2 (t_{l_1} \ldots t_j)^{-1/2}\right). \end{aligned} \tag{5.24}$$

The functions V_1 and u_3 depend on t_γ' via the factors $t_\gamma' \varkappa_\gamma{}^2 \pi_{\gamma'}(\boldsymbol{\tau})$ where $\pi_\gamma(\boldsymbol{\tau}) = \prod \tau_{l'}$ and the product runs over l' with $\gamma \subset \gamma_{l'-1}$ and $\gamma^* \cap \gamma_{l'-1} = \emptyset$. As to the function V_2, it depends on t_γ' at $\gamma \notin \mathcal{F}_\alpha{}^c$ via these factors only, and the parameters t_γ' at $\gamma \in \mathcal{F}_\alpha{}^c$ enter in the combinations $\varkappa_\gamma \sqrt{t_\gamma'}$. The analysis of convergence of $R_{\mathcal{F},\mathcal{N}} G_\Delta$ with respect to $\tau_{l'}$ (IR convergence) and t_l ($l = 1, 2, \ldots, l_1 - 1$; UV convergence) was, in fact, performed in Subsection 5.2. It shows either non-negativity of the relevant powers or the presence of the functions φ which provide the convergence in $\tau_{l'}$. Therefore, one has to establish only the convergence with respect to other variables, t_l at $l = l_1, \ldots, j$.

To evaluate the asymptotic behaviour of $R_{\mathcal{F},\mathcal{N}} G_\Delta$ at $\varrho \to 0$ let us use the Proposition 5 of ref. [13]: the integral

$$F^{\mp}(\varrho) = \int_0^1 \prod_{l=1}^{j} t_l^{N_l}\, dt_l \exp\left\{\mp \frac{i\varrho^2}{t_{l_1} \ldots t_j}\right\}$$

at $N_l \geqq 0$ for $l \leqq l_1$ defines, at $\varrho > 0$, an infinitely differentiable function that admits the representation

$$F^{\mp}(\varrho) = \sum_{l=l_1}^{j} P_l^{\mp}(\ln \varrho^2)\, \varrho^{2(N_l+1)} + F_{\mp}^{(0)}(\varrho^2)$$

with $P_l^{\mp}$ polynomials and $F_{\mp}^{(0)} \in C^\infty$ at $\varrho \geqq 0$. The derivatives in $\varkappa_\gamma$ $(\gamma \in \mathcal{F} \cup \Sigma_{\mathcal{N}})$ may operate on different factors u_1, u_2, u_3 in (5.24). If $\partial/\partial\varkappa_\gamma$ operates on u_1 then the power N_l at $l = l_1, \ldots, j$ becomes $N_l - 1$. The proposition mentioned above shows that the power of ϱ is decreased by 2. However, this is cancelled due to the factor ϱ^2 that appears because of the same differentiation of u_1. Analogous remark take place in respect of the factor u_2. Therefore, when evaluating the behaviour of $R_{\mathcal{F},\mathcal{N}}G_\Delta$ at $\varrho \to 0$, one suffices to consider only differentiation of u_3. Thus, the problem reduces to the power counting of $t_l(l_1 \leqq l \leqq j)$ that would be performed if the UV convergence analysis was considered at $\varrho = 0$. This power counting rests on relations (5.13) — (5.15) and gives

$$N_l \geqq \sum_{\substack{1 \leqq i \leqq k \\ \sigma(\gamma^i) \leqq l}} \frac{1}{2}\,(a^{\gamma^i} - a^{\gamma^{i-1}}) = \frac{1}{2}\,(N - a_1 - a_2)$$

as $a^{\gamma^i} = N - a_1 - a_2$ for any $i = 1, \ldots, k$.

With the help of Proposition 5 of ref. [13] and its consequences one obtains that the nonanalytic part of the asymptotic expansion of the integral $R_{\mathcal{F},\mathcal{N}}G_\Delta$ at $\varrho \to 0$ begins with $\varrho^{N-a_1-a_2+1}P(\ln \varrho^2)$. This behaviour is sufficient to provide the desired order of the remainder because the action of $1 - \mathcal{M}_0$ reduces to $(1 - \mathcal{M}_\zeta^{N-a_1-a_2})\, R_\xi G_\Gamma(\zeta\varrho\xi_1, \zeta\varrho\xi_2, \boldsymbol{p})$.

6. Analytic renormalization

1. In massless theories with dimensionless coupling contants the analytic renormalization (AR) [14] removes both UV and IR divergences [9]. Let us now describe a modification of AR that makes it possible to eliminate the divergences from theories almost without IR divergences.

Consider FA $G_\Gamma(\boldsymbol{p})$ corresponding to a graph from a theory of the given class. Associate now with any SE-subgraph $\gamma \in \Sigma$ the subtracting operator of the zero order $M_\gamma^0 \equiv M_\gamma$ defined by (2.4) at $n = 0$. Let $\mathcal{F}$ be the evaluator involved in the definition of AR [14]. The following statement holds.

Theorem 6.1. 1) The operator $R = \mathcal{F} \prod_{\gamma \in \Sigma} (1 - M_\gamma)$ removes both UV and IR divergences. 2) The operator R performs R-operation.

2. Let us describe the main steps of a proof of the theorem. We use the α-representation of analytically regularized FA $G_\Gamma(\boldsymbol{p}, \boldsymbol{\lambda})$: the regularization is introduced in (5.1) by the factor $\prod_l (\mu^2\alpha_l)^{\lambda_l}$. Then, as in Sect. 5, we turn to the dimensionless α_l, perform the decomposition into sectors, and introduce the sector variables by (5.3).

To analyse the sector contribution $G_\Delta(\boldsymbol{p}, \boldsymbol{\lambda})$ let us rewrite the product $\prod_{\gamma \in \Sigma} (1 - M_\gamma)$ in a convenient form. Represent Σ as $\Sigma = \Sigma_{\mathcal{N}} \cup \bar{\Sigma}_{\mathcal{N}}$ (the definition of $\Sigma_{\mathcal{N}}$ may be found in Sect. 5). Then

$$\prod_{\gamma \in \Sigma} (1 - M_\gamma) = \prod_{\gamma \in \Sigma_{\mathcal{N}}} (1 - M_\gamma) \sum_{F \in \{F\}} \prod_{\gamma' \in F} (-M_{\gamma'}). \tag{6.1}$$

Here the sum runs over the forests composed of the elements of the set $\bar{\Sigma}_{\mathcal{N}}$, including the empty forest. Consider the typical term from the sum over F. Let us choose mixed representation (5.2) according to the prescriptions of ref. [15] with $\mathcal{L}_a = \{1, 2, \ldots, j\} \cup \mathcal{L}(F)$. The exponent of the representation turns to be a continuous function of the sector variables due to the choice of the representation. One easily verifies that the action of $X_F = \prod_{\gamma \in F} M_\gamma$ provides the same function in the exponent of the representation as in the exponent associated with FA corresponding to the reduced graph Γ/F. Furthermore, the factor $Z^r(\boldsymbol{x}, \boldsymbol{p}, \alpha, \beta)$ transforms, due to X_F, into the function that gives the contribution to the powers of the sector variables which is not less than the contribution of Z^r itself. Finally, the factor $(A^r)^{-2} \equiv (D_{\hat{\Gamma}^2})^{-2}$ becomes $\left(\prod_{\gamma \in F} D_{\gamma/\underline{\gamma}} A^r_{\Gamma/F}\right)^{-2}$ where $\underline{\gamma}$ is the union of the elements of F strictly included in γ, and $A^r_{\Gamma/F} \equiv D_{\hat{\Gamma}^2/F}$.

To analyze the convergence one should apply factorization properties of the forms $D_{\gamma/\underline{\gamma}}$ and $A^r_{\Gamma/F}$. Any form is represented as a product of the variables t_l, $\tau_{l'}$ and the parameters $\varkappa = \{\varkappa_\gamma \mid \gamma \in F\}$ in some powers and infinitely differentiable functions which depend on $\varkappa_\gamma$ via the combinations

$$\sigma_\gamma^2 = \varkappa_\gamma^2 \prod_{l \in \Lambda_1(\gamma)} t_l \prod_{l' \in \Lambda_2(\gamma)} \tau_{l'}. \tag{6.2}$$

Here $\Lambda_1(\gamma)$ is the set of $l \leqq j$ such that γ is a connectivity component of γ_l; $\Lambda_2(\gamma)$ is the set of $l' \geqq j + 1$ such that γ is a connectivity component of $\gamma_{l'-1}$.

As a result the contribution to $G_\Delta(\boldsymbol{p}, \lambda)$ from the typical term of (6.1) becomes (up to non-essential factor)

$$\int_0^1 dt\, d\tau \prod_l t_l^{N_l+\nu_l} \prod_{l'} \tau_{l'}^{N'_{l'}-\bar{\nu}_{l'}} \prod_{\gamma \in \Sigma_{\mathcal{N}}} (1 - M_\gamma)$$
$$\times \varkappa_\gamma^{n_\gamma} Q(\boldsymbol{x}, \boldsymbol{p}, \boldsymbol{t}, \boldsymbol{\tau}, \boldsymbol{\sigma}) \exp\left(-i \sum_l m_l^2 \alpha_l(\boldsymbol{t}, \boldsymbol{\tau})\right) \tag{6.3}$$

with $\nu_l = \sum_{n=l}^{l} \lambda_n$, $\bar{\nu}_{l'} = \sum_{n=l'}^{L} \lambda_n$, $Q \in C^\infty$.

The factorization properties give

$$N_l \geqq -[1/2\omega(\gamma_l)] - 2 \sum_{\gamma \in F} \eta(\gamma_l, \gamma) - 1 \tag{6.4}$$

$$N'_{l'} \geqq -\Omega(\Gamma \setminus \gamma_{l'-1}) - 2 \sum_{\gamma \in F} \eta(\hat{\gamma}_{l'-1}, \gamma) - 1. \tag{6.5}$$

Here $\eta(\gamma', \gamma)$ if $\mathcal{L}(\gamma^*) \subset \mathcal{L}(\gamma')$ and two lines of γ^* are not connected with each other by a path in $\gamma \cap \gamma'$. Otherwise $\eta(\gamma', \gamma) = 0$. Note that $\eta(\gamma_l, \gamma) = 0$ for any $\gamma \in \Sigma_{\mathcal{N}}$ and l. The definition of the IR index Ω may be found in Sect. 5. Furthermore,

$$n_\gamma \geqq -4\mathfrak{N}(\gamma) - 2[1/2a(\gamma)]. \tag{6.6}$$

Now one should consider the action of $\prod_{\gamma \in \Sigma_{\mathcal{N}}} (1 - M_\gamma)$. Let us represent $1 - M_\gamma$ as $\int_0^1 d\varkappa_\gamma \times \partial/\partial\varkappa_\gamma\, \varkappa^{4\mathfrak{N}(\gamma)+a(\gamma)}$ (compare (5.12)). Thus, the action of $1 - M_\gamma$ reduces to the differentiation of the first order in $\varkappa_\gamma$. With the use of (6.2) expression (6.3) takes the form

$$\prod_{\gamma \in \Sigma_{\mathcal{N}}} \int_0^1 d\varkappa_\gamma \int_0^1 dt\, d\tau \prod_{l=1}^{j} t_l^{\bar{N}_l+\nu_l} \prod_{l'=j+1}^{L} \tau_{l'}^{\bar{N}'_{l'}-\bar{\nu}_{l'}}$$
$$\times Q_1(\boldsymbol{p}, \boldsymbol{x}, \boldsymbol{t}, \boldsymbol{\tau}, \boldsymbol{\sigma}) \exp\left\{-i \sum_l m_l^2 \alpha_l(\boldsymbol{t}, \boldsymbol{\tau})\right\} \tag{6.7}$$

with

$$\overline{N}_l \geqq N_l + \sum_{\gamma \in \Sigma_{\mathcal{N}}: l \in \Lambda_1(\gamma)} 1, \tag{6.8}$$

$$\overline{N}'_{l'} \geqq N'_{l'} + \sum_{\gamma \in \Sigma_{\mathcal{N}}: l' \in \Lambda_2(\gamma)} 1. \tag{6.9}$$

Let us show that the action of M_γ does not generate additional singularities in λ. These are determined by the powers N_l of t_l. If M_γ are absent then the variables t_l occur in the powers $N_l \geqq -[1/2\omega(\gamma_l)] - 1$. At first sight formula (6.4) may imply that the degrees are possibly decreased. However, since $\eta(\gamma_l, \gamma) \neq 0$ only for 1 PI subgraphs γ_l (see the definition of Σ and γ^*) then one easily shows that there is a subgraph γ^1 such that $N_l \geqq -[1/2\omega(\gamma^1)] - 1$. Moreover, (6.8) shows that the degrees of t_l are, in fact, increased due to $\prod_{\gamma \in \Sigma_{\mathcal{N}}} (1 - M_\gamma)$. Thus, some singularities in λ disappear. Hence, the function $\prod_{\gamma \in \Sigma} (1 - M_\gamma)\, G_\Gamma(p, \lambda)$ is not "more singular" in λ that $G_\Gamma(p, \lambda)$ itself.

Let us now establish the absence of IR singularities at the point $\lambda = 0$. To do this, one suffices to achieve the inequality $\overline{N}'_{l'} > -1$ when the subgraph $\gamma_{l'-1}$ includes all the massive lines. In the case of the theories considered rhs of (6.5) may be negative only if $\gamma_{l'-1}$ contains a subgraph from $\Sigma_{\mathcal{N}}$ as a connectivity component. Any such component contributes the value -1 to $N'_{l'}$. However, due to the definition of $\Sigma_{\mathcal{N}}$, this contribution is cancelled completely in (6.9) when $\overline{N}'_{l'}$ is considered. As to the contribution of the terms with η, they are unessential: to show this one must represent Ω as a difference of UV indices and repeat the arguments in respect of the contribution of similar terms into N_l. Thus, $\overline{N}'_{l'} > -1$ for any "dangerous" l', and this implies the IR convergence because for other l' the convergence in $\tau_{l'}$ is provided by the massive exponent, irrespective of $\overline{N}'_{l'}$.

3. In ref. [15] (see also the preceding paper) it was proved that for any graph Γ without massless tadpoles there exists a region of regularizing parameters λ where the corresponding FA $G_\Gamma(p, \lambda) \in \mathcal{S}'(R^{4n})$ is an analytic function (this result is a generalization of Speer theorem [16] for the case of arbitrary external momenta). The analysis of this section has shown that for diagrams of the class under consideration the "partially" renormalized FA $G_\Gamma{}^{(0)}(p, \lambda) = \prod_{\gamma \in \Sigma} (1 - M_\gamma)\, G_\Gamma(p, \lambda)$ has an analyticity domain that is not less than the domain of $G_\Gamma(p, \lambda)$ itself. Now we let the evaluator $\mathcal{F}$ [14] operate on $G_\Gamma{}^{(0)}$. As a result we have the analytically renormalized FA RG_Γ. To show that the operation R, indeed, performs the renormalization it suffices to use the results of ref. [17] where an analogous statement was proved for massive FA. The techniqual difference between the present paper and ref. [17] concerns different types of sectors and the corresponding sector variables. However, in proving the second part of the theorem this difference is unessential. In fact, one must repeat the scheme of the proof of ref. [17].

Finally, we note that to establish the absence of IR poles corresponding to subgraphs with massive lines it is possible to use the integration by parts over the variables $\tau_{l'}$ [15] (see also Subsection 5.2).

We are grateful to M. C. Polivanov for support.

References

[1] Bogoliubov, N. N., Parasiuk, O. S., Acta Math. **97**, 227 (1957).
[2] Hepp, K., Commun. Math. Phys. **2**, 301 (1966).
[3] Zimmermann, W., Ann. Phys. **77**, 536 (1973).
[4] Anikin, S. A., Polivanov, M. C., Zavialov, O. I., Fortschr. Phys. **27**, 527 (1977).
[5] Gomes, M., Lowenstein, J. H., Zimmermann, W., Commun. Math. Phys. **39**, 81 (1974).
[6] Lowenstein, J. H., Zimmermann, W., Nucl. Phys. **B 86**, 77 (1975); Commun. Math. Phys. **44** 73 (1975).
[7] Anikin, S. A., Karchev, N. I., Zavialov, O. I., Theor. Math. Phys. **44**, 751 (1980); **46**, 1 (1981). Anikin, S. A., Bordag, M., Zavialov, O. I., Teor. Mat. Fiz. **56**, 163 (1983).
[8] Breitenlohner, P., Maison, D., Commun. Math. Phys. **52**, 39; 55 (1977).
[9] Anikin, S. A., Smirnov, V. A., Theor. Math. Phys. **51**, N 1 (1982).
[10] Wilson, K. G., Phys. Rev. **179, 1499** (1969).
[11] Zimmermann, W., Ann. Phys. **77**, 570 (1973).
[12] Anikin, S. A., Zavialov, O. I., Theor. Math. Phys. **26**, 105 (1976).
[13] Anikin, S. A., Zavialov, O. I., Theor. Math. Phys. **27**, 585 (1976). Anikin, S. A., Zavialov, O. I., Ann. Phys. **116**, 135 (1978). Anikin, S. A., Theor. Math. Phys. **52**, 829 (1982). Anikin, S. A., Karchev, N. I., Zavialov, O. I., Theor. Math. Phys. **38**, 291 (1979).
[14] Speer, E. R., J. Math. Phys. **9**, 1404 (1968).
[15] Smirnov, V. A., Theor. Math. Phys. **44**, 761 (1980); **46**, 132 (1981).
[16] Speer, E. R., Ann. Inst. H. Poincare **23**, 1 (1975).
[17] Speer, E. R., Commun. Math. Phys. **23**, 23 (1971); **25**, 226 (1972).
[18] 't Hooft, G., Nucl. Phys. **B 61**, 455 (1973).
[19] Tkachov, F. V., Phys. Lett. **124 B**, 212 (1983).
[20] Chetyrkin, K. G., Phys. Lett. **126 B**, 371 (1983).

The journal "Fortschritte der Physik/Progress of Physics" is devoted to the theoretical and experimental study of the fundamental constituents of matter and their interactions, so to elementary particle physics, classical and quantum field theory, theory of graviation, thermodynamics and statistics, nuclear physics, laser physics, and plasma physics.
The articles should have in general review character.

Manuscripts should be addressed to the Editor-in-Chief of the journal

Prof. Dr. F. Kaschluhn
Sektion Physik der Humboldt-Universität zu Berlin
DDR-1086 Berlin, PSF 1297

or to editors or to members of the Advisory Board.

Terms of subscription for the journal "Fortschritte der Physik/Process of Physics"

Orders can be sent

- *in the GDR:* to the Deutsche Post, Zentralvertrieb des PZV (B), 7930 Herzberg/Elster, or to the Akademie-Verlag Berlin, DDR-1086 Berlin, Leipziger Str. 3–4, PF-Nr. 1233;
- *in the other socialist countries:* to a book-shop for foreign language literature or to the competent news-distributing agency;
- *in the FRG and Berlin (West):* to a book-shop or to the wholesale distributing agency Kunst und Wissen, Erich Bieber OHG, Wilhelmstr. 4–6, D-7000 Stuttgart 1;
- *in the other Western European Countries:* to Kunst und Wissen, Erich Bieber GmbH, Dufourstr. 51, CH-8008 Zürich;
- *in other countries:* to the international book- and journal-selling trade, to Buchexport, Volkseigener Außenhandelsbetrieb der DDR, Postfach 160, DDR-7010 Leipzig, or to the Akademie-Verlag Berlin, DDR-1086 Berlin, Leipziger Str. 3–4, PF-Nr. 1233.

Zeitschrift „Fortschritte der Physik"

Herausgeber: Prof. Dr. F. Kaschluhn, Prof. A. Lösche, Prof. Dr. R. Rompe.
Verlag: Akademie-Verlag Berlin, DDR-1086 Berlin, Leipziger Str. 3–4;
Fernruf: 2236201 und 2236229; Telex-Nr.: 114420; Bank: Staatsbank der DDR, Berlin, Konto-Nr.: 6836-26-20712.

Chefredakteur: Dr. Lutz Rothkirch
Anschrift der Redaktion: Sektion Physik der Humboldt-Universität zu Berlin, DDR-1040 Berlin, Hessische Straße 2.

Veröffentlicht unter der Lizenznummer 1324 des Presseamtes beim Vorsitzenden des Ministerrates der Deutschen Demokratischen Republik.

Gesamtherstellung: VEB Druckhaus „Maxim Gorki".
DDR-7400 Altenburg, Carl-von-Ossietzky-Straße 30/31.

Erscheinungsweise: Die Zeitschrift „Fortschritte der Physik" erscheint monatlich. Die 12 Hefte eines Jahres bilden einen Band. Bezugspreis je Band 216,– DM zuzüglich Versandspesen. Preis je Heft 18,– DM. Der gültige Jahresbezugspreis für die DDR ist der Postzeitungsliste zu entnehmen.

Bestellnummer dieses Heftes: 1027/33/9.

AN (EDV) 57618

01000

Contents of the following issues:

W. Lucha: Proton Decay in Grand Unified Theorie

V. I. Komarov, H. Müller, S. Tesch: Cluster Excitation in Hadron-Nucleus Collisions at Medium and High Energies

J. J. Haubold, A. M. Mathai: The Maxwell-Boltzmannian Approach to the Nuclear Reaction Rate Theory

H. Dorn: Renormalization of Path Ordered Phase Factors and Related Hadron Operators in Gauge Field Theories

T. I. Kopaleishvili: Quantum Field Theory Approach to the Pion-Two-Nucleon Interaction Problem

J. Maharana: Canonical Structure and Quantization of Grassmann Variables

W.-D. Nowak: Review of Deep Inelastic Charged Lepton Scattering

E. Sezgin: The Spectrum of $D = 11$ Supergravity via Harmonic Expansion on $S^4 \times S^7$

Progress of Physics is indexed
in Current Contents/Physical, Chemical & Earth Sciences.

FSC
www.fsc.org
MIX
Papier | Fördert
gute Waldnutzung
FSC® C083411

Zeitfracht Medien GmbH
Ferdinand-Jühlke-Straße 7
99095 Erfurt, Deutschland
produktsicherheit@kolibri360.de